NEW ZEALAND'S
LOST HERITAGE

First published in 2013 by New Holland Publishers (NZ) Ltd
Auckland • Sydney • London • Cape Town

www.newhollandpublishers.co.nz

218 Lake Road, Northcote, Auckland 0627, New Zealand
Unit 1, 66 Gibbes Street, Chatswood, NSW 2067, Australia
The Chandlery, Unit 114, 50 Westminster Bridge Road, London, SE1 7QY, United Kingdom
Wembley Square, First Floor, Solan Road, Gardens, Cape Town 8001, South Africa

Copyright © 2013 in text: Richard Wolfe
Copyright © 2013 in photographs: individual suppliers as credited on page 192
Copyright © 2013 New Holland Publishers (NZ) Ltd
Richard Wolfe has asserted his right to be identified as the author of this work.

Publishing manager: Christine Thomson
Editor: Geoff Walker
Design: Book Design Ltd, www.bookdesign.co.nz

Front cover images: Seacliff Lunatic Asylum (top), Hiona, Maungapohatu (bottom)
Back cover images, from left to right: T.J. Edmonds Ltd Factory, Lyttelton Timeball Station, Victoria Arcade

National Library of New Zealand Cataloguing-in-Publication Data

Wolfe, Richard.
New Zealand's lost heritage : the stories behind our forgotten
landmarks / Richard Wolfe.
Includes bibliographical references and index.
ISBN 978-1-86966-387-2
1. Lost architecture—New Zealand. 2. Historic buildings—New
Zealand—3. New Zealand—History. I. Title.
993—dc 23

10 9 8 7 6 5 4 3 2

Printed in China by Toppan Leefung Printing Ltd, on paper sourced from sustainable forests.

While every effort has been made to contact copyright holders of material used in this book, the publisher will correct any omissions or errors in future editions of the work.

All rights reserved. No part of this publication may be reproduced, stored in a retrieval system, or transmitted in any form or by any means, electronic, mechanical, photocopying, recording or otherwise, without the prior permission of the publishers and copyright holders.

NEW ZEALAND'S
LOST HERITAGE

THE STORIES BEHIND OUR FORGOTTEN LANDMARKS

RICHARD WOLFE

CONTENTS

Introduction 7
1. A Complete Fortress: **Ruapekapeka** 16
2. That Monstrous Bungle: **Admiralty House** 26
3. The Limbo of Lost Things: **Coolangatta** 34
4. A Fine Pile of Buildings: **His Majesty's Theatre and Arcade** 40
5. The Most Beautiful View: **Kilbryde** 52
6. Sentinel over Auckland: **Partington's Mill** 60
7. Ecclesiastical and Homelike: **St Paul's** 70
8. 'The Finest Erection in Auckland': **Victoria Arcade** 78
9. Mystic Signs at Maungapohatu: **Hiona** 88
10. A Lantern in the Town: **The Round House** 96
11. Primitive Grandeur: **St John's Cathedral** 102
12. A Unique Blend: **Rangiatea** 110
13. Wellington Gothic: **Parliament Buildings** 118
14. A Hall from where Laws Emanate: **Nelson Provincial Government Building** 126
15. Home of the Rising Sun: **T.J. Edmonds Ltd Factory** 134
16. Assisting Those at Sea: **Lyttelton Timeball Station** 142
17. In the Scottish Baronial Manner: **Seacliff Lunatic Asylum** 150
18. Handsome and Appropriate: **The 1865 New Zealand Exhibition Building** 158
19. The Most Complete in the Country: **Invercargill Post Office** 166
20. A Great Sanitary Agent: **Dee Street Hospital** 174

References 182
Index 188
Picture Credits 192

INTRODUCTION

A melancholy aspect

Fire, earthquake and, in particular, the wrecker's ball have taken a heavy toll on the buildings of New Zealand. This book presents the stories of a broad selection of some which no longer exist, or at least not in their original form. The loss of a small number of these may be attributed to acts of God, but the vast majority were removed by deliberate acts of man. Two of these buildings were victims of earthquakes and three were destroyed by fire – at least one of which was lit intentionally – but three-quarters of this number have been summarily demolished, creating an irreplaceable gap in the country's architectural record and collective cultural memory.

An early example of this country's lost heritage is Kawakawa chief Kawiti's pa, Ruapekapeka, in Northland, whose wooden huts and stockading were burned by British troops after the battle of 11 January 1846. In ordering this action, Colonel Henry Despard was no doubt thinking of the safety of British settlers rather than the preservation of a remarkable example of military engineering. However, he did decide to leave intact the pa's extensive trenches and earthworks, remnants of which can still be admired today.

Little sentiment

Conflicts with Maori from the mid-1840s encouraged British settlers to abandon the northern part of the country for the relative safety of Auckland. Expansion of that rapidly growing settlement was hampered by several prominent headlands, including Point Britomart, which was systematically cut down from the early 1870s. English historian

LEFT: *Workmen lined up along Auckland's Greys Avenue following the completion of the Salvation Army Congress Hall, or Citadel, in 1928.* PREVIOUS: *Interior view of Rangiatea showing two of the three 12-metre-high totara pillars which support the massive 26.2 metre-long tauhu, or ridgepole. Also visible are the additional supports for the beams supporting the rafters, installed for structural reasons in the early 1900s.*

James Froude was in Auckland at the time and observed the removal of the point and its buildings, among them the first major Gothic Revival church erected in New Zealand. Presciently, Froude commented that colonials appeared to have little sentiment when it came to matters of heritage. Thirty years later, the need for new roads led to the destruction of one of Auckland's most remarkable – and redundant – residences, Admiralty House, while the combined demands of railways and shipping would also bring about the demise of Kilbryde, the grand home of the city's founding father, Sir John Logan Campbell.

With the prevalence of timber construction, fire was a constant hazard in the settlements of early New Zealand. Auckland, for example, suffered several catastrophic fires, including one which broke out on a Saturday night in September 1873. It destroyed 54 houses, shops and places of business, levelling a vast section of Queen Street. As reported in a local newspaper, 'So swift and so absolute was the progress of destruction that scarcely a piece of timber remains, and the whole of the property above-mentioned disappeared in little more than two hours. On Sunday morning, Queen Street presented a very melancholy aspect.'[1]

A spectacular fire also consumed the colony's Parliament Buildings in Wellington in late 1907, while the first half of the new century saw the loss by demolition of some unique examples of local architecture, from Rua Kenana's extraordinary temple, Hiona, at Maungapohatu, and Dunedin's 1865 New Zealand Exhibition Building, to New Plymouth's Round House and Invercargill's post office. This period was also marked by the tragedy of the 1931 Napier earthquake, which resulted in the destruction of St John's Cathedral.

A country 'too new'
Nine years later, the 1840–1940 New Zealand Centennial celebrations and the gift to the nation by Governor-General Lord Bledisloe of the Treaty House at Waitangi contributed to raising the country's awareness of its own history. Such interests were further heightened by the controversial destruction in 1950 of Auckland's landmark Partington's Mill. Two years later this event was a factor behind the presentation to Parliament of a private member's National Historic Places Trust Bill, whose stated purpose was to 'mobilise local and national interest in identifying, retaining, and suitably marking … the various sites of buildings, institutions, battlegrounds, Maori pas, and other historic places of interest to Maori and pakeha …' During discussions in the House it was suggested that New Zealanders thought of their country as being too new to have anything in the way of antiquities, and had an 'easy-going' attitude towards the past.

Even so, it was acknowledged as remarkable that in 1920 Auckland had managed to preserve Sir John Logan Campbell's first home, Acacia Cottage, the oldest timber dwelling in the city, by relocating it from O'Connell Street to the safety of Cornwall

INTRODUCTION

Park. And to illustrate the parlous state of local heritage, it was pointed out that there were now only two survivors of New Zealand's nine original provincial council buildings, a number which would be further reduced with the demolition of the Nelson complex in 1969.[2]

Our dubious record

It was apparent that New Zealand lagged behind most other Commonwealth countries in the matter of preserving records and marking historical sites. And although the National Historic Places Trust Bill itself did not proceed, the Government assumed responsibility for the necessary legislation. The resulting Historic Places Act of 1954 established the National Historic Places Trust, which first met the following year (and changed its name to the New Zealand Historic Places Trust in 1963).[3]

By way of comparison, the heritage movement in Britain got under way in the late 19th century with various societies which could be seen as ancestors of the National Trust, which was established in 1895. But if New Zealand has a dubious record, Britain has also allowed some major lapses. Captain James Cook occupies a unique position in both countries' histories, but the two houses he occupied in East London have gone. One, on Mile End Road, was demolished as recently as 1959, despite having a London City Council commemorative plaque affixed to it in 1907. The location of the other house, on The Highway, Wapping, is now marked by a blue plaque which records its one-time resident as 'The ablest and most renowned Navigator this or any country hath produced'. The cottage in which Cook was born in North Yorkshire was also demolished in 1786, but at least his boyhood home has survived; after being sold in 1933, it was shipped to Australia and re-erected in Melbourne's Fitzroy Gardens.[4]

British heritage experienced more intensive threats in the 1940s in the form of German bombing, and from the comprehensive urban redevelopment that followed. Among the major casualties of the latter was the glorious 1869 Columbia Market in Bethnal Green, demolished in 1958 and described as 'one of London's most grievous architectural losses'. Four years later another noble landmark fell, the massive 1838 Greek Doric propylaeum that was the Euston Arch, gateway to the railway station of the same name. Its demolition did much to boost the growth of the conservation movement, and galvanise popular reaction against 'the institutional philistinism which characterised the post-war period'.[5]

By the second half of the 20th century, railway stations around the world were coming under threat. The destruction in 1963 of New York's magnificent Beaux Arts-style Penn Station, built in 1908, prompted international outrage and was the catalyst for a local architectural preservation movement. London had lost its Euston Arch, and a few years later it took a spirited campaign to save the nearby and exuberantly Gothic St Pancras station from a similar fate. Auckland's own Railway Station, completed in 1931, became redundant in 2003 with the opening of the Britomart Transport Centre in the former Central Post Office, but it survives following conversion to student accommodation.

Across the Tasman, Australia's interest in its colonial legacy was stimulated in the 1890s when fire destroyed early buildings in downtown Sydney. A decade later, the removal of sections of The Rocks on the waterfront following an outbreak of bubonic plague led to the realisation that local heritage was fast disappearing. But it was the threat that post-Second World War development posed to Sydney's architectural heritage that led to the formation of the National Trust in 1947. It was modelled on its English parent, with an initial focus on the nation's stately homes, and for most of the 1950s and 1960s it was almost the only such advocate for preservation.[6]

Auckland's oldest house

Caught between the loss of Partington's Mill and the passing of this country's first protective legislation was what was believed to be Auckland's oldest house. Owners of the 109-year-old Kitchener Street property had decided on demolition, while descendants of the original builder suggested the house be preserved and shifted to another site. The city council and owners agreed to look into the matter, but nothing came of it. And so, by late July 1951 the house had become 'a rubble of dust and timber'. Built soon after Auckland was founded in September 1840, the two-roomed house had been extended but thereafter remained substantially unaltered, and when demolished its original (handsawn) timbers were said to be in excellent condition, with no sign of borer.[7]

If London's Columbia Market had a New Zealand equivalent it might conceivably be Auckland's Victoria Arcade, pulled down in 1978. Described as one of that city's 'most valuable cultural landmarks'[8] and by conservation architect Jeremy Salmond as 'an astonishing building' and 'a symphony in red brick',[9] it had a strong connection with local artists, providing studios and headquarters for the Auckland Society of Arts. But this once striking building had been allowed to deteriorate, and then suffered the added indignity of being replaced by a particularly ugly office block.

Sign of things to come

Ominously, the passing of the Arcade was a sign of things to come, for the deregulated economy of the mid-1980s unleashed market forces which would change the face of the nation's inner cities. The best known of such losses was His Majesty's Theatre and Arcade, another early gem in Auckland's architectural crown. This complex had also

RIGHT: *The north-west corner of Shortland Street showing the prominent corner tower of the Victoria Arcade (left). The Post & Telegraph Office building was demolished shortly after this photograph was taken in 1925.*

been wilfully neglected and was at the mercy of egregious developers and a compliant city council, until finally succumbing to a pre-emptive strike by a demolition squad.

Two other significant losses in Auckland at this time, not detailed in this selection, were the Salvation Army Citadel (or Congress Hall) and the Chapel of the Little Sisters of the Poor. When the first of these was completed in 1928 it was said to have assisted in the beautification of 'a once neglected street', Greys Avenue.[10] This delightfully ornate two-storey building was topped by a tower with Moorish corner turrets which would not have been out of place in the Civic Theatre nearby. But a sit-in by protesters in early 1990 could not prevent its demolition, to make way for the construction of Mayoral Drive.

A decade later Auckland allowed the departure of another architectural treasure in the form of the Chapel of the Little Sisters of the Poor, in Tweed Street, Ponsonby. In a curious twist, the force behind this loss was the Little Sisters themselves, who had decided to redevelop their property. Their Romanesque chapel had been designed and completed in 1908 by architect Thomas Mahoney, also responsible for St Benedict's Church in Newton and the Auckland Customhouse. The Historic Places Trust notified its intention to seek protection for the chapel, but it was too late, and the site is now occupied by a pale version of Mahoney's original.[11]

In 1964 *The New Zealand Herald* lamented the demolition of Auckland's early buildings, among them the city's old hospital (whose four towers had stood 'squarely against the sky'), the Waverley Hotel and, predictably, Partington's Mill, and 'the grandest of all houses', Kilbryde. Although grateful for what had survived – the Auckland Art Gallery, Library, Central Post Office and Supreme Court – the writer understandably wondered where they might be in 20 years' time.[12] Remarkably, perhaps, all four made it to 1984, and are still with us today, although some serious adjustments have been necessary. The old Library building has been absorbed by the Art Gallery (which itself has recently undergone a major redevelopment), the Post Office has been transformed into the Britomart Transport Centre, and the Supreme Court is now part of Metropolis, a 40-storey residential/hotel skyscraper.

In addition to many losses there have been some uncomfortably close calls. As outrageous as it may seem now, demolition of Auckland's Civic Theatre was considered as early as 1940, when it was barely 11 years old, and again in the 1980s. Fortunately the 'Mighty Civic' survived those threats, and was restored to its former glory in the late 1990s. But buildings have had to suffer various indignities in order to survive. In 1982, for example, Auckland's 94-year-old St Mary's Church was wrenched from its historic site and hauled across to the other side of Parnell Road. Far worse was that decade's vogue for 'façadism', whereby only the front of a building was retained, and juxtaposed with a new structure rearing up menacingly behind it. Such a superficial deference insulted the integrity of the original building which, apart from reasonable and sensitive modification, was either worth saving or it wasn't. Kirkaldies in Wellington and the Clarendon Hotel (since demolished) in Christchurch are examples of this 'pseudo-conservation', while Auckland has what architectural historian Peter Shaw describes as the 'hopelessly vulgarised' Queen's Head Tavern in Queen Street.[13]

Earthquake risks

Recent earthquakes in Christchurch have provided a graphic reminder of the potential seismic threat to the buildings of New Zealand. Of that city's 300 heritage buildings, 200 were lost, either irrevocably destroyed by the earthquakes themselves or subsequently demolished for reasons of instability.[14] The New Zealand Historic Places Trust attempted to save those that it felt posed no risk of collapse, and questioned claims that others were uneconomic to repair. However, its powers were usurped by the Canterbury Earthquake Recovery Act (CERA), and the list of buildings demolished includes St John the Baptist Church, the Regent Theatre and Canterbury Library.[15]

At the time of writing, the fate of two of the city's most prominent buildings hung in the balance. The Catholic Cathedral of the Blessed Sacrament, completed in 1905, was closed after the 4 September 2010 earthquake, but a subsequent shock caused the collapse of the two front bell towers and destabilised the dome, which was then removed. The decision as to whether the building would be restored or demolished was yet to be decided, and was dependent on engineering reports.[16]

The Christchurch Cathedral (the Anglican Cathedral Church of Christ), which dates from 1864, also suffered in the February 2011 earthquake, which destroyed the spire and part of the tower and severely damaged the remaining structure. It was decided that the building would be deconsecrated, and in March 2012 Bishop Victoria Matthews announced it would be demolished, a decision supported by CERA and church groups but strongly opposed by engineers, individuals (including the city's resident Wizard) and a Restore Christ Church Cathedral group. In the meantime, Japanese architect Shigeru Ban was working on the design of a temporary transitional cathedral, to accommodate some 700 people. The A-frame structure, based on 64 waterproofed and fireproofed cardboard tubes, was erected on Latimer Square. After several construction delays, the world's first cardboard cathedral was officially opened with a dedication service on 15 August 2013.

In early April 2013 the Anglican Diocese (of Christchurch) and the Church Property Trustees announced three new design options for the damaged Christchurch Cathedral: the complete repair and rebuilding of the existing building; replacement of part of the building; and the full demolition of the existing cathedral and its replacement with a new contemporary structure. At the same time, the highly controversial issue of the demolition of the building was the subject of ongoing court action.[17]

Also badly damaged in 2011 were the Canterbury Provincial Council Buildings, which date from 1859. In particular, the collapse of the 1865 stone council chamber was considered by some to have been the greatest single heritage loss of the February earthquake. Since the demolition of the provincial council buildings in Nelson in 1969, the much admired Canterbury buildings were the only surviving example of a complete provincial government complex. For architectural historian Ian Lochhead, resurrecting the damaged buildings would demonstrate local resilience, restore a unique link with the colonial past, and reflect the confidence and prosperity of 1860s Canterbury.[18]

New earthquake standards

While Wellington had some early requirements for buildings to improve their seismic performance, the first national design standards were introduced in 1935, a consequence of the Napier earthquake.[19] These were later updated, and in December 2012, in response to a Canterbury Earthquake Royal Commission, the Government proposed more stringent measures. These included the seismic assessment of all commercial and high-rise, multi-unit buildings in the country (some 193,000 properties), and the requirement that all those that were not upgraded to withstand a moderate-sized earthquake within 10 years would be demolished. Owners would need to strengthen existing properties to a safe level, being 34 per cent of the design level of a new building.

Such a proposal would have a major impact on Auckland's character areas such as Ponsonby and Karangahape Roads, as well as such centres as Whanganui, Dunedin and Invercargill, which have large numbers of heritage structures.[20] There would be massive implications, not least for owners who would be required to foot the bill for all strengthening work. There would be little likelihood of a financial return on what might be a very expensive investment, a prospect which might encourage owners to walk away from the property, or to intentionally defer maintenance. The latter course of inaction would likely lead to irreversible dilapidation, with 'demolition by neglect' then an inevitable outcome. One wonders if just such a fate awaits Auckland's 1928 St James Theatre. The last of the Queen Street theatres built for live shows, it was closed following a fire in 2007 and has languished ever since.

The prospect of property ownership was an incentive from the earliest days of organised settlement in New Zealand. It came to be considered a natural right, encouraging an attitude expressed in Maurice Shadbolt's 1978 short story 'The People Before': 'I'd bend my head to no man. And you know what the secret of that is, boy? Land of your own. You're independent, boy. You can say no to the world. That's if you got your own little kingdom.'[21]

Further to this is the belief that ownership of 'kingdoms' confers absolute rights, irrespective of whether the property has a heritage value that might be of local – or national – significance. The 2006 loss of Coolangatta in Remuera, Auckland, was a case in point.

The city council voted against protecting the house, believing that if it did so it could be required to compensate an out-of-pocket owner who was bent on demolition. Coolangatta may be the most recent addition to the list of grand New Zealand residences that are no more, along with Admiralty House and Kilbryde. Many had names as impressive as their extravagant architecture – Longbeach (Ashburton), Cashmere (Christchurch), Siberia (Manawatu) and Longwood (Wairarapa); and some 70 examples of such houses were the subject of Terence E.R. Hodgson's 1978 book *Fire & Decay*.[22]

A ban on demolition?

Help may be at hand for the more typical houses of New Zealand. Following some highly publicised cases, in late November 2012 the Auckland Council advised that it was considering a region-wide ban on the demolition of heritage and character homes unless owners could prove they were beyond repair. Based on a Brisbane model, the new rules would apply to all pre-Second World War properties, representing 23,344 houses in existing character zones, and thousands elsewhere.[23] While this would offer a stay of execution for vulnerable properties, it would need to not encourage fossilisation; old houses require (sensitive) adaptation in order to remain practicable. At the same time, talk of the need for high-density city living will surely put pressure on older housing stock.

With the release in March 2013 of the draft Auckland Unitary Plan, the need to provide for future population growth (one million over the next 30 years) had massive implications for certain areas. A timely perspective on some of the city's more vulnerable old buildings – as found along Karangahape and Ponsonby Roads – was provided by an exhibition of photographs by Allan McDonald. He described Auckland's architectural heritage preservation record as 'appalling', whereas Dunedin maintained a 'fiercely protective' attitude towards its own comparatively rich stock of surviving buildings. McDonald's exhibition 'Walking in the City' followed the footsteps of Eugène Atget, whose camera had documented the vulnerable side of Paris a century or so earlier.[24]

Consideration of the architectural merits of lost buildings should not overlook the need to ensure high standards of design at the outset. Buildings can also acquire significance through association, although that didn't help those inhabited by Captain Cook. Further, those built without posterity in mind, and lacking any notable connections, can also be worthy of recognition; a humble worker's cottage, for example, can illuminate an early chapter in our social history. There have been suggestions that we need to erect 'iconic' buildings, ignoring the logic that only the passage of time will enable us to confer such status. We lament our lost heritage, but future generations will at least be able to enjoy that which is spared the wrecker's ball today. And while we may not be able to anticipate future attitudes, perhaps we should prepare for the prospect of a piece of façadism receiving heritage protection, as a worthy exemplar of a period in our past.

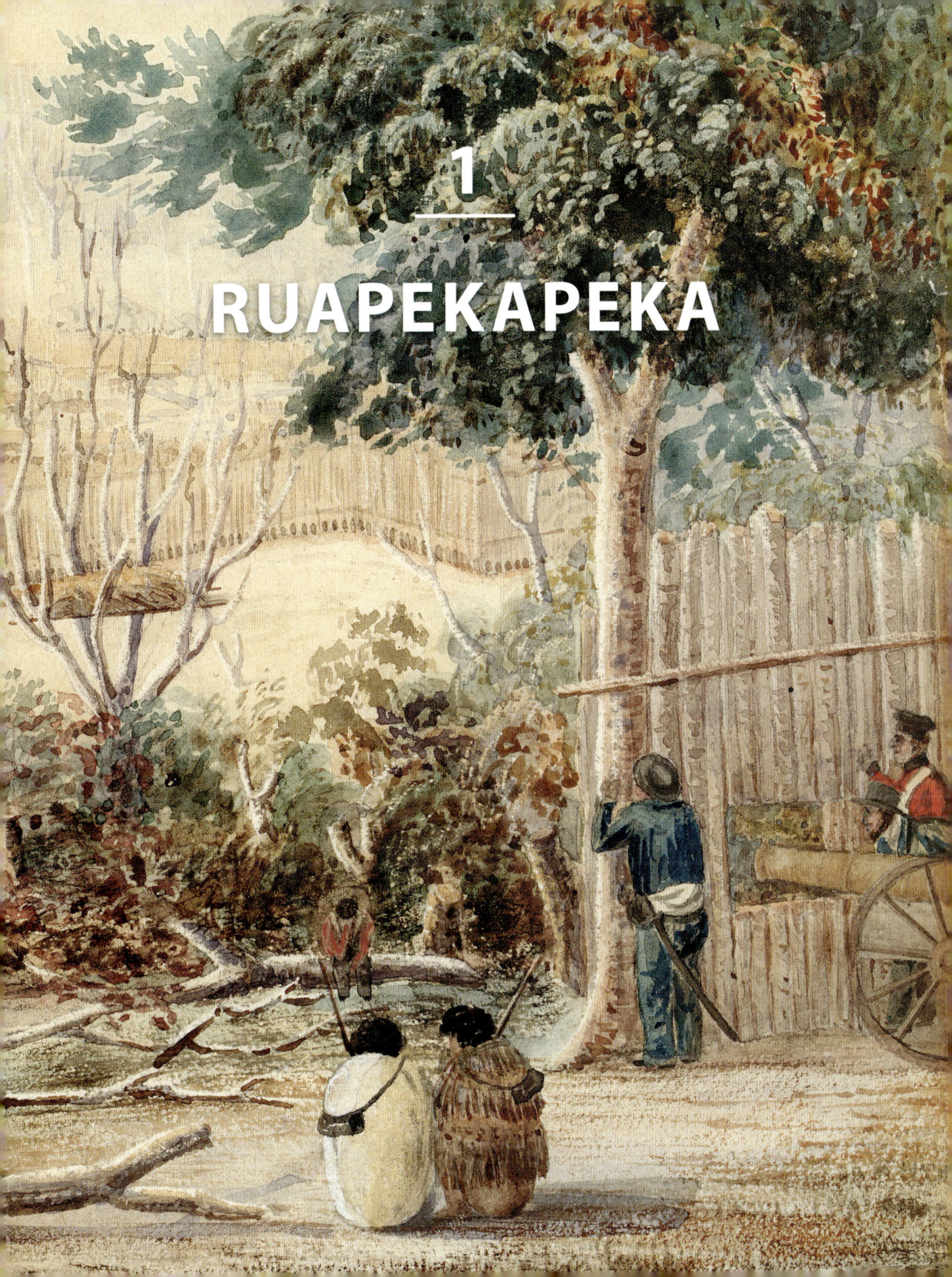

1
RUAPEKAPEKA

A COMPLETE FORTRESS:
RUAPEKAPEKA

AN ISOLATED HILLTOP south of the Bay of Islands is the site of the last battle in the first phase of the New Zealand Wars, which took place between 1845 and 1872. The fortified pa of Ruapekapeka was a massive undertaking, although nothing now remains of the timber palisading which offered stiff resistance to British artillery fire and set the pattern for Maori defence works that followed. However, sufficient of the earthworks have survived to give a visitor a sense of the scale of the original stockade, and the challenge it posed for an attacking force. It was the extent and nature of these defences that now qualify Ruapekapeka as a site of both national and international importance.[1]

In July 1844 Ngapuhi leader Hone Heke Pokai made the first of several attacks on the symbol of British authority, the flagpole at Kororareka. On 11 March 1845 he and Kawakawa chief Kawiti struck again, a blow which resulted in the destruction of much of the town. The European population was evacuated, and there were fears Auckland might now be under threat. A British force was sent north, and a series of conflicts followed. One of these was at Ohaeawai, near Lake Omapere, a fortified position built by Kawiti to withstand British artillery fire. In fact, a week-long bombardment had little effect, and the engagement proved a severe set-back for the British troops, who suffered heavy losses.

Preparing for a showdown
Governor FitzRoy's handling of the worsening situation was found wanting, and he was recalled and replaced by the then governor of South Australia, George Grey.[2] He arrived in Auckland in November 1845 and shortly ordered all available forces to be sent north. The troops were stationed at Kororareka, where officers occupied the house of Bishop Pompallier and soldiers pitched tents where the town had once stood. Grey made contact with the rebellious chiefs, but they were reportedly 'very little inclined to be quiescent'.[3] Further conflict seemed inevitable, and Kawiti prepared for the showdown by building a new fortress, Ruapekapeka, said to be stronger than Ohaeawai.

ABOVE: *A recent view of a cannon and the remnants of the earthworks at Ruapekapeka.* PREVIOUS: *A watercolour painting by British Officer Cyprian Bridge of the final assault on Ruapekapeka, on 11 January 1846.*

Ruapekapeka was a large pa, some 150 m by 70 m, and situated on 'an eminence' about 40 km from Kororareka. It was surrounded by rows of 3.5 m-high palisades consisting of trunks of puriri – a hard timber known to British settlers as 'iron wood' – and split timber lashed together and reinforced with layers of flax to absorb the impact of cannon balls. The palisades were positioned to allow defenders to fire in all directions, while trenches were designed to provide protection against raking fire and the ricochet of cannon balls.[4]

Inside the pa was a system of shot- and bomb-proof underground refuges, or rua, to give shelter from bombardment. Ruapekapeka, 'the bats' nest', was so-named because these underground chambers had narrow entrances like those in the natural habitats of the native bat. Also scattered within the pa were felled tree trunks and raised mounds, to both provide cover in case of attack and to impede the progress of an assault. And beyond the defences, intersecting roads had been cut through the bush leading to the pa to enable Kawiti's men to fire on approaching troops.[5]

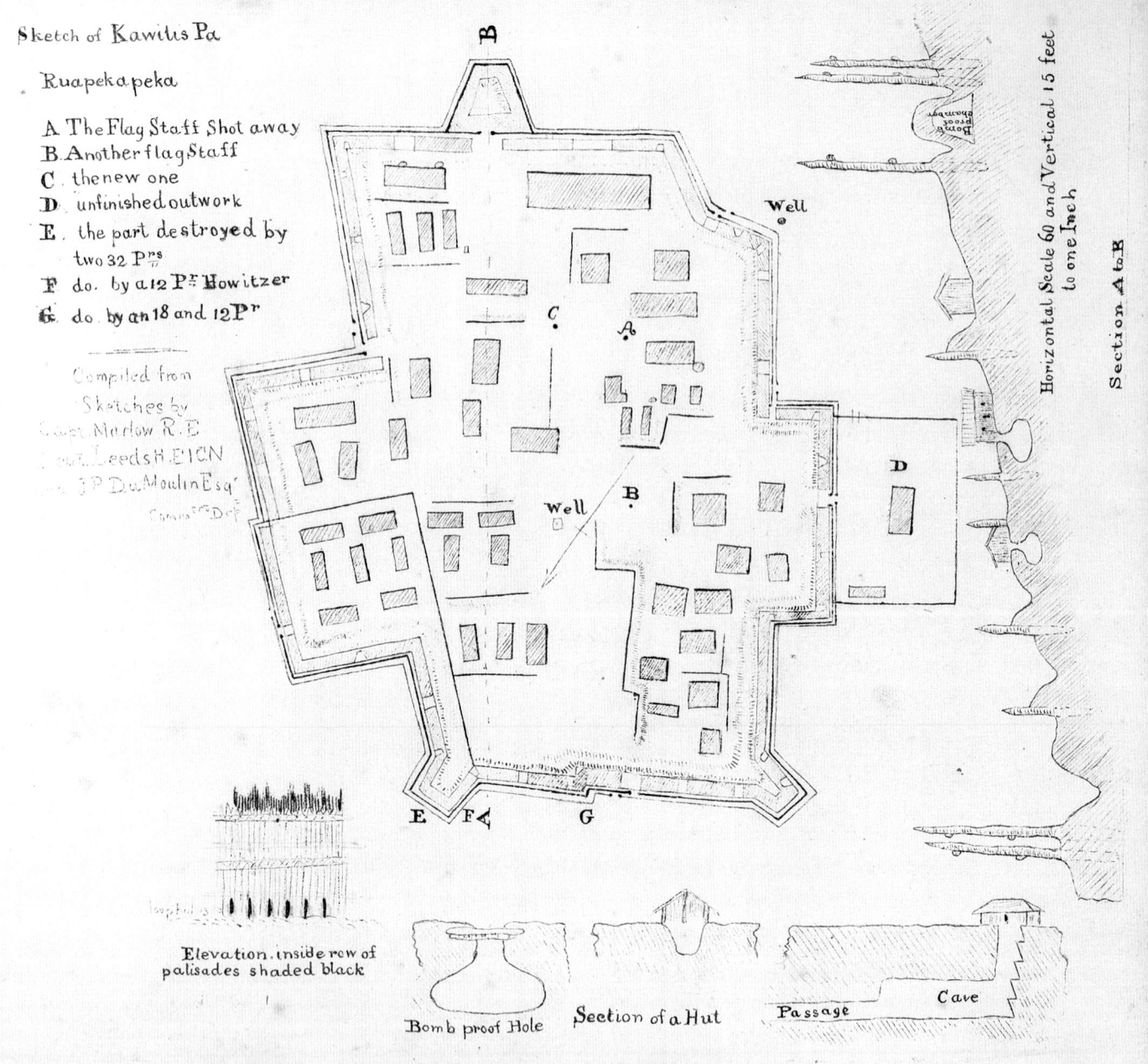

Drawing of Ruapekapeka, showing sections destroyed by the bombardment, and details of fortifications and underground shelters. This image, probably lithographed in London, c.1846, was by an unknown artist and compiled from sketches by Capt. Marlow, Lieut. Leeds and J.P. du Moulin.

Historian James Cowan referred to the 'soldierly genius' of Kawiti in selecting the position of Ruapekapeka. It was situated on the end of a narrow neck of land, nearly 335 m above sea level on the northern face of the Tapuaeharuru Range, and was remote and difficult to approach. It provided a commanding view of the surrounding countryside, including the shipping off Kororareka, so British movements could be anticipated.[6]

The advance begins

On 8 December 1845 the British force under Colonel Henry Despard began its advance on Ruapekapeka. Supporting and supply ships sailed up the Kawakawa River, and from there boats went as far as high tide would permit. Thereafter lay some 24 km of 'roadless hills, forests, swamps, and streams'. Some 30 tonnes of supplies and artillery needed to be shifted, including three naval 32-pounders, one 18-pounder, two 12-pounder howitzers, one 6-pounder brass gun, four mortars and two rocket-tubes. Today a road running from Kawakawa east of State Highway 1 largely follows the route taken by Despard's men. It was a major undertaking, requiring the felling of bush, the construction of tracks and bridges, and the use of block and tackle to haul the guns over rough ground and up steep hills.[7]

> **TRENCH WARFARE**
>
> Traditional Maori warfare was based on hand-to-hand combat, but new strategies were needed to counter the introduction to this country of muskets and artillery. Trench warfare is usually associated with the Western Front in Europe during the First World War, but in fact such field works date back to Roman times. In his 1980 book *The New Zealand Wars*, and in a subsequent 1998 television documentary, historian James Belich claimed that by developing such fortifications Maori invented a new system of warfare. This claim has been disputed by some military historians, but there is no doubt that defensive works as such constructed at Ruapekapeka represented an ingenious response to the threat posed by British firepower.

The total number involved in the attack on Ruapekapeka was 68 officers and 1100 men, as well as 450 friendly 'native allies' under five chiefs including Tamati Waka Nene and Patuone. This force was over three times the size of that in Ruapekapeka which, with the arrival of Hone Heke and his men, numbered some 500.[8] The British troops established a main camp about 700 m from the pa, and two forward gun batteries at a closer range. Guns were brought up by means of horses, bullock teams and manpower, and axemen cleared the bush immediately in front of the advanced gun positions, exposing much of Ruapekapeka to the British fire. The first contact between the two sides took place on 1 January 1846 when a small party from the pa engaged in tree-to-tree fighting with 'friendly Maori'.[9]

What was described by James Cowan as the 'grand bombardment' began on the morning of 10 January. 'Every gun spoke' as shot was hurled against the front of the palisade and lobbed into the pa itself. Kawiti had two pieces of artillery in the pa – a 12-pounder carronade and a 4-pounder – and men who could operate them, but early in the conflict the 12-pounder took a direct hit on the muzzle from an 18-pound shot and was destroyed. A 'storm of shot and shell' was kept up all day, and some of the pa's palisade posts, nearly 6 m high and more than 30 cm thick, were either shattered or knocked out of the ground by the impact of the cannon balls.

By the afternoon the stockade had been breached, and Despard would have launched a storming party — as he had done with dire consequences at Ohaewae — but was dissuaded from doing so by Governor Grey, who was present during the whole operation. The barrage continued into the night, with guns and rockets being fired regularly to prevent the defenders of Ruapaekapeka from carrying out running repairs to their stockade.[10]

The next day, Sunday, the barrage continued, but there was no return musket fire from the pa. A party of Maori scouts, followed by troops, crept through a breach in the defences, and found the pa largely deserted. Kawiti had remained inside with some of his followers, but most of the garrison had left and were sheltering behind the rear earthworks. Kawiti now attempted to regain control of the pa, but his men were outnumbered by the soldiers pouring into the stockade. The invaders also rushed out of the rear of the pa and were now drawn into the bush by retreating Maori, falling into a cunning ambush laid by Kawiti. Here they became easy targets, and most of the British casualties at Ruapekapeka — 12 killed and 30 wounded — occurred in the fighting that took place behind the pa.[11]

TAMATI WAKA NENE

Tamati Waka Nene was born around the 1780s into a family connected to many of the leading chiefs in the north of New Zealand. He first distinguished himself in battle in the early 1800s and, recognising the advantages of a Pakeha presence, sought to preserve the peace during this turbulent period. In 1839 he was baptised and took the name Tamati Waka, after Thomas Walker, an English settler associated with the Church Missionary Society. Following the destruction of Kororareka in 1845, Nene was the leading Maori chief who took the Government side against Heke and Kawiti. He was a man of great mana, and held in high regard by Pakeha, and when he died in 1871 he was buried in the Christ Church Cemetery, Russell.

Outcome was a draw

One theory as to why Ruapekapeka had been abandoned was that the Christian Maori had left the pa to hold a Sunday prayer service. Kawiti, a non-Christian, had stayed within the stockade. But the battle gave neither side the victory it sought.[12] While the capture of such a well-constructed pa was a significant achievement, and may have represented a tactical victory for the British, the outcome was more of a draw. Heke and Kawiti survived, and with their forces largely intact, and the terms of the peace settlement that followed suggested they had achieved a strategic victory.[13]

A COMPLETE FORTRESS: RUAPEKAPEKA

Ruapekapeka pa, on the hill in the distance, under bombardment on 10 January 1846. This watercolour was originally attributed to Major Cyprian Bridge, but is now believed to have been painted by Sgt. John Williams.

After the battle, British troops set fire to the huts and stockading at Ruapekapeka. But the earthworks and the trench system were so sizeable that Despard decided to leave them intact, and then marched his troops back to the Bay of Islands.[14] In his despatch, Despard admitted to the skill and originality of the Maori in the building of Ruapekapeka: 'The extraordinary strength of this place, particularly in its interior defences, far exceeded any idea I could have formed of it. Every hut was a complete fortress in itself, being strongly stockaded all round with heavy timbers sunk deep into the ground, and placed close to each other, few of them being less than one foot in diameter, and many considerably more, besides having a strong embankment thrown up behind them. Each hut had also a deep excavation close to it, forming a complete bomb proof shelter, and sufficiently large to contain several people, where at night they were completely sheltered from both shot and shell.'[15]

The Auckland newspaper *The New Zealander* found the result of the battle at Ruapekapeka a 'source of great satisfaction', but felt that the circumstances under which the pa was taken did not justify the 'lengthened, pompous, commendatory despatches' of Colonel Despard. The paper, which aimed to represent the interests of the average settler and Maori,[16] considered Despard's reference to 'the capture of a fortress, of extraordinary strength, by assault, and nobly defended by a brave and determined enemy' an exaggeration, and so offered 'a plain, unvarnished narrative of the facts'. It advised that a party of friendly Maori had entered the breached defences of the pa unopposed, and the following Government troops rushed in before Kawiti's men – who were outside the pa engaged in their *karakia* (worship) – were able to re-enter.

No enemy within
The New Zealander referred to this 'anomalous *assault* on an enemy's fortification, which had no enemy within it' and an official brigade order which claimed the nature of the capture of the pa proved the 'intrepidity and gallantry of all concerned'. Colonel Depard had identified no less than 21 individuals who 'conspicuously distinguished themselves', whereupon *The New Zealander* wondered whether drummers and fifers, who received no mention, might feel deprived of 'their laurels in this assault'.[17]

In 1922 James Cowan wrote that when he visited Ruapekapeka he saw the charred remains of palisade posts, and crumbling parapets covered with fern, flax and the hebe koromiko. He observed a post some 3.7 m high and 36 cm in diameter which was white with age and charred from the fire of 1846. He imagined such defences presenting 'a formidable face to the attacking force', the height from the bottom of the outer ditch to the top of the now fern-covered maioro, or earth wall, being 4.6 m. Cowan entered a rua, a 'comparatively wide chamber' with moss-covered floor some 2 m below ground

> **JAMES COWAN**
> James Cowan (1870-1941) was one of New Zealand's most widely read non-fiction writers during the first half of the 20th century. He was born at East Tamaki, Auckland, and began his professional career as a reporter on the *Auckland Star*. He became a prolific writer with a strong interest in Maori and Pakeha history, with over 30 books and hundreds of articles to his credit. His first book was published in 1901, and two years later he became a journalist for the recently established Department of Tourist and Health Resorts. He later became a freelance writer. The book for which he is best known is the two-volume *The New Zealand Wars: A History of the Maori Campaigns and the Pioneering Period*, which was published in 1922–23 and remained until recently the definitive account of those years.

and roofed over with logs and earth. He recognised the convenience and safety of the subterranean shelter during a bombardment, and described the whole pa as being 'pitted with burrow-like ruas' while its extensive trench system would still 'conceal a little army'.[18]

Kawiti's last stand

Ruapekapeka was Kawiti's finest and strongest fighting pa. It was also his last stand and the final battle in the 'War in the North'. In addition to its location – inland in the middle of nowhere, and requiring attackers to travel a considerable distance over rugged country – it was far better equipped to deal with cannon fire than traditional pa. It was not the first pa to employ trenches or strong palisades, but its adaptations of these made it one of the most effective against musket and artillery fire. As a result, this system was adopted and developed by others around the country during the next 30 years. And while the earlier Maori success at Ohaewai had dispensed with the sense of European superiority, Ruapekapeka further demonstrated the sophistication of Maori military techniques.

Although the timber palisades have long gone, the remnants of the earthworks qualify Ruapekapeka as one of the best-preserved pa of the New Zealand Wars. Further, according to historian Nigel Prickett, it 'illustrates the conceptual thinking and engineering skill' that went into such structures.[19] The nearby bush contains some surviving large puriri trees, providing a sense of the size and effectiveness of the stockade which resisted British artillery here some 170 years ago.

2
ADMIRALTY HOUSE

THAT MONSTROUS BUNGLE:
ADMIRALTY HOUSE

ONE OF THE MOST EXTRAVAGANT residences Auckland has ever known had the additional and unique distinction of never being used for its intended purpose. There were claims that it should never have been built in the first place, and it lived out its short life as a boarding house (at least it was a 'fashionable' one). It also managed to survive a fire, but in the end it stood in the way of progress and was demolished to make way for a new road.

In the early 19th century New Zealand's defence depended on occasional visits by ships of the Royal Navy based in Australian waters. In response to demands by Governor Grey, in September 1863 the British Admiralty directed the commodore of the Australia station, William Wiseman, to make his squadron headquarters for a period in Auckland. Shortly after his arrival he commanded a flotilla of gunboats on the Waikato River, and was also responsible for the establishment of a permanent naval base at Devonport, Auckland.

The need to accommodate Wiseman in Auckland led to the construction of the first Admiralty House, in Short Street, Official Bay. It was built on an Admiralty reserve and paid for by the Provincial Government. But by 1866 Wiseman was no longer commodore of the Australia station, and his departure raised concerns about responsibility for the house. A member of the Provincial Council asked whether such free quarters should in fact be provided, and whether the house should be rented out when not occupied by a naval commander.[1]

Seven years later there was talk of pulling Admiralty House down. While it was considered a pleasant enough building with commanding views, it was said to have been badly built and in an unattractive neighbourhood.[2] However, a suggestion that the current commodore might make a lengthy stay in Auckland if the house was reburbished led to the necessary work being carried out.[3] The following year the house was readied for Wiseman's replacement, Rear-Admiral Wilson, who was visiting with the squadron. Grounds were laid out with the assistance of prison labour, the improvements including a

ABOVE: *Admiralty House in Jermyn Street in 1903, looking south east from Emily Place. The state of the grounds and the boarded-up entrance (right) suggest this photograph was taken prior to the premises becoming the Glenalvon boarding house.* PREVIOUS: *Admiralty House, c.1906, with its commanding view of the Waitemata Harbour and Rangitoto.*

---

privet hedge to relieve the 'baldness' of a paling fence on the sea frontage. The property also included a dilapidated fowlhouse ('with its inevitable concomitants') inexplicably placed beside the front verandah, and it was felt this 'hideous excrescence' should be removed before the arrival of the Rear-Admiral.[4]

A source of pride

By 1897 Auckland was the official Imperial naval station for New Zealand. The presence of British ships in the Waitemata Harbour was considered a source of pride to citizens, and a reminder that they held 'a not unimportant place in the great Imperial System'.[5] But the now aging Admiralty House was no longer considered suitable for its purpose. The Auckland Harbour Board proposed to Premier Richard John Seddon that the Government sell the present house (for an anticipated £1500–2000) and erect a new one on another site (to be provided free of cost by the Auckland Harbour Board and Auckland City Council) at a cost of £3000, and furnish it at a cost of £1000. The harbour board was providing 'every facility within its means for the convenience of Her Majesty's navy', and hoped the shortfall of £2000 would be met by the Government. Further, along with the Auckland City Council, the board was also giving up its share of a valuable piece of land.[6]

The proposal went forward, and during a reading of the subsequent Admiralty House Bill, in 1898, Parliament heard it was the desire of the citizens of Auckland, and of the colony generally, that the commander of the squadron visit their shores as often as convenient. It was claimed the Bill would ensure his stay in New Zealand waters was 'more pleasant and more attractive'. The legislation required the Government to hand over the existing Admiralty House and site, valued at £2500, plus an additional £1000 towards the estimated cost of the new premises of £5000.[7]

The architect of the second Admiralty House was Charles Le Neve Arnold, whose other important buildings in Auckland would shortly include the Great Northern Brewery and Colonial Sugar Refinery Office (1903) and later, in partnership with R. Atkinson Abbot, Auckland Grammar School (1913) and King's College Memorial Chapel (1922).[8] Admiralty House, situated on the slopes below present-day Anzac Avenue, was a grand multi-storeyed edifice, distinguished by a complex plan surmounted by steep roofs and two towers with conical tops. Its prominent porch, projecting gables and turrets, were features of the Queen Anne style, particularly popular with American mansions in the late 19th century, while it was also said to have been based on villas in the French Riviera and Switzerland.[9] Naturally the house had a ballroom and a billiards room, and, while seven bedrooms was not an unusual number at the time, having six lavatories was considered excessive.[10]

THE 1898 ADMIRALTY HOUSE BILL

This Bill was to provide for a suitable site and to erect and furnish a new house and grounds for the use of the Commander in Chief of the Australasian Naval Station. The Government would grant £1000 towards the project, while the existing Admiralty House and grounds would be vested in the Auckland Harbour Board. And in the event of the house not being needed by the Naval Commander, the board would also be required to finance – out of its own general funds – its use by 'any other distinguished Imperial or Colonial officer'. The Admiralty House Bill was read and passed in late October 1898, at a time when Parliament was also considering such other diverse pieces of legislation as the Noxious Weeds Bill, the Lunatic Act Amendment Bill, and the Slander of Women Bill.

A white elephant?

But barely had building of Admiralty House begun in 1901 when there were accusations of it being already a 'white elephant'. The *Observer* labelled the harbour board's 'precipitancy' in accepting a tender for the job as 'nothing short of folly'. And apart from not having the approval of the Lords of the British Admiralty, the house was in a 'wholly unsuitable position'.[11] Thoughtful Aucklanders were quick to offer alternative uses for the building, their suggestions including a technical college, veterans' home, casualty ward and refuge for the homeless.[12]

The building was completed – the exterior, that is – and in late February 1903 arrangements were being made for a public welcome to Auckland for the new commodore of the naval station, Admiral Fanshawe.[13] But the harbour board was about to receive a bombshell, for Fanshawe advised that he would not be making use of Admiralty House. His duties made it impossible for him to spend more than a few days a year at Auckland, and as the senior naval officer of the New Zealand division he felt that his headquarters should be on board his ship.[14]

The board was now forced to consider its options, the most obvious one being to lease the house. But the Government would not allow it to do so, for the reason that the house had been authorised by a special Act of Parliament, which would need to be repealed or amended before it could be either let or sold. In the light of this, and echoing an earlier poultry reference, the *Observer* noted the harbour board's 'fatuous and impractical extravagances of the last few years are coming home to roost'.[15]

The 'Admiral-less House'

The impasse prompted the board to send another deputation to Wellington, to inform the Premier that the empty house was costing the board £554 a year. Seddon was 'disgusted' with what had happened, and felt there should be a dissolution of the partnership between the board and the Government. As for alternatives, he knew of no government department which could find a use for Admiralty House,[16] which was now being referred to as 'Admiral-less House'.[17]

The 1898 Act required the Auckland Harbour Board, on providing Admiralty House ready for the reception of 'the elusive Admiral', to receive from the Government the old house and site, in addition to a subsidy of £1000. But the latter conditions could not be met because the new house was not yet furnished, the board having decided it would be imprudent to invest another £1000 without the certainty of some return.[18] The fiasco stimulated further strong comment, with Admiralty House being labelled 'that monstrous bungle' which had already cost the port of Auckland nearly £10,000, plus interest and maintenance, while there did not appear to be 'the remotest probability' that it would ever be used as intended by an admiral. One local newspaper suggested,

Admiralty House and the Waitamata Harbour beyond viewed from the Grand Hotel in Princes Street. The two buildings on the right, on the corner of Eden Crescent, are the Auckland Museum (from 1876–1929) and its caretaker's cottage.

---

somewhat cynically, that it was a consolation for Auckland to know that it now had a house which an admiral might live in, 'if he felt so inclined'. All the same, it was 'a useless luxury'.[19]

The harbour board was reminded that it knew before Admiralty House was even built that the acting-admiral had decided against living – even temporarily – in Auckland, and that naval authorities believed that the right place of residence for an admiral on duty was his ship. But perhaps the location of the house had something to do with the problem, for while it had sweeping views of the Waitemata, the immediate foreground was dominated by the less picturesque prospect of railway tracks, coal depots and timber yards. Harbour board members blamed the Government, which had adopted the plans for the house, and the naval authorities, who had chosen the site. The board had hoped the house would be similar to one in Hobart, a kind of 'naval club' that would attract ships of the squadron, which would in turn bring business to the city.[20]

Mrs Scherff's boarding house

The impasse was broken when the harbour board was given power to deal with the house. It was offered on a 14-year lease with the tenant bound by certain conditions, one being the painting of the exterior with two coats at least once every five years.[21] The premises now became the Glenalvon boarding house, under the management of Mrs Scherff.[22] But on the afternoon of 11 November 1906 fire broke out in the kitchen and a portion of the building was destroyed. Had it not been for the prompt action of the City Fire Brigade, the whole premises were likely to have been razed to the ground. About 60 persons were resident at the time, but the only casualties were two firemen injured following the failure of a fire escape.[23]

A month later architect Charles Arnold submitted plans for reinstating the Glenalvon, at an estimated cost of £1756,[24] while the harbour board reduced Mrs Scherff's monthly rent by £10 during the period the destroyed portion of the building was not in use.[25] The Glenalvon was soon back in business, and the following year was described by an Otago newspaper as 'a fashionable boarding house'.[26] But its history had not been forgotten; the *Auckland Star* continued to criticise the harbour board's 'reckless expenditure', and its list of 'relics of the past' included 'the house built for the Admiral who never came'.[27]

In late 1911 plans were announced for five rooms to be added to the Glenalvon.[28] But four years later, following the decision to redevelop Jermyn Street as part of Auckland's new eastern outlet (to be known as Anzac Avenue), the house stood in the way of progress. Not surprisingly perhaps, the harbour board was happy to relieve itself of the property, so it was sold to the city council and demolished. In November 1915 Admiralty House's timbers ('first class heart of kauri'), galvanised iron, doors, windows, staircases and gas fittings went on sale on the premises.[29] The Glenalvon, however, lived on, for Mrs Scherff (who had previously operated the same business in Symonds Street) now secured new premises known as Bella Vista nearby in Waterloo Quadrant.[30]

In 1931 the late Admiral House was referred to as having been a 'baronial structure'.[31] Its architect, Charles Arnold, died a quarter of a century later at the age of 100, thereby far outliving that ill-fated property which survived for only some 14 years.

3
COOLANGATTA

THE LIMBO OF LOST THINGS:
COOLANGATTA

IN 1991 A PROPERTY on Auckland's Remuera ridge was described as one of the city's 'most magnificent houses' and 'most elegant architectural sights'.[1] But just 15 years later this house, Coolangatta, was hastily reduced to a pile of rubble. Its lack of protection and subsequent demolition once again drew attention to the vulnerability of the city's heritage, despite suggestions that under a new mayor such acts of urban vandalism would not be repeated. The loss of this gracious 1913 mansion, considered 'Auckland's most admired residence',[2] was equated with two other well-known gaps in its architectural heritage, Partington's Mill and His Majesty's Theatre and Arcade.[3]

Coolangatta had stood at 464 Remuera Road for 93 years, and during that time had been owned by just two families. Its name came from that given to an estate established south of Sydney, Australia, in the 1820s by Alexander Berry, a Scottish ship's surgeon, captain, explorer and merchant. After his death the property passed to other members of his family, eventually resulting in a bequest which enabled Jessie Foster to buy the Remuera section on which she and her husband, Alfred, built Coolangatta. She paid £1100 for the property, and the estimated contract price for the house in late 1912 was £1000.[4]

Return to traditional ideals

Coolangatta was designed by architect F. (Frederick) Noel Bamford. His design for the house reflected the influence of the Arts and Crafts movement which had developed in Britain from the 1850s. It sought a return to traditional ideals of honest craftsmanship and simple forms, and encouraged a vernacular approach to architecture using ordinary and local materials. Arts and Crafts houses were distinguished by their simple solid forms, wide porches and steep roofs, and use of brick and tile. The style was brought to New Zealand around the end of the 19th century, and became the preferred choice of architecturally designed houses for the next four decades, standing out from the standard villas and bungalows.[5]

ABOVE: *The north (rear) face of Coolangatta, showing the verandah which afforded a view of Waitemata Harbour and Rangitoto. The simple form of the house with a steep roof was characteristic of the Arts and Crafts style.* PREVIOUS: *The elegant street frontage of Coolangatta (1913–2006).*

A long, sweeping driveway led to Coolangatta, with its entrance porch copied from a house by English architect Edwin Lutyens.[6] The exterior of the house was of lime-washed double-brick construction, with pale-blue Welsh slate roofing and woodwork of mostly oiled kauri. Rustic casement windows evoked the 18th century, and elements of the interior looked back to English country houses of earlier times. Features of the interior were sliding doors between the drawing and dining rooms, a smoking room, a sewing room and a maid's wing. A long narrow verandah ran along the rear of the house, while the top floor enjoyed a view of the harbour and Rangitoto. The carefully crafted house was an adaptation of overseas design ideas to local conditions, and as a result it has been suggested that the house could have existed nowhere else but New Zealand.[7]

In 1954 Jessie Foster sold the property for £15,000 to Morton Coutts. His grandfather, Joseph Kuhtze, had arrived in New Zealand in 1867 from Cologne, Germany, and was a pioneer of the local brewing industry, while his son William founded the Waitemata Brewery, which became part of Dominion Breweries. Morton Coutts followed the family tradition; as an inveterate inventor he discovered a new and faster method of brewing beer, the 'continuous fermentation process'. It was patented worldwide in 1956, and within four years was responsible for 85 per cent of all the beer brewed in New Zealand.[8]

One of the few survivors
Following Morton Coutts's death in 2004 and his widow's move to a retirement home, Coolangatta was put on the market. It was advertised as being zoned Residential 7, which allowed intense development. Because of the size of the section, 12 units would be

THE ARCHITECT

New Plymouth-born F. (Frederick) Noel Bamford (1881–1952) studied under Auckland architect Edward Bartley during the construction (from 1902) of St Matthew's Church, and then worked for a period in the office of leading English architect Edwin Lutyens. He returned to New Zealand in 1906, entering a partnership with A.P. Hector Pierce. Their designs included Bishopscourt, for the Anglican Bishop Neligan, in St Stephens Avenue, and buildings for the 1913–14 Auckland Exhibition in the Domain, of which only the Tea Kiosk survives.[9]

permitted.[10] Coolangatta had made a big impression on many people and was one of the few surviving mansions on Remuera Road. It was one of the best examples of architecture of its era, and one of the few with both section and building fabric in an original state.[11]

It was widely assumed that the house enjoyed some form of protection under Auckland City's heritage rules, but in fact it had none. As a result, in 2006 it was recommended that the property be scheduled and recognised as an important building with a unique architectural style, and so be given protection under the Auckland City Council's District Plan. But council members argued that scheduling the house could result in considerable compensation costs, for which there was no apparent budget. At issue was the difference in economic value between a potential 12-unit site and one encumbered with a protected house. The city services manager recommended that the city council not protect the house, on the grounds that if it were to do so the Environment Court could require it to purchase the scheduled property.

Mindful that Coolangatta was already on the market, the majority of council – along with the Coutts family – opposed the scheduling because it was felt the removal of rights conferred by existing zoning would affect the legitimate interests of the owner. And so in May 2006 the council voted against the recommendation. Further, it was decided that details of how individual members had voted on the issue be kept secret.[12]

By early December 2006 it was apparent that Coolangatta may have been sold. On the morning of 9 December a demolition team with two large excavators arrived on site and rapidly destroyed the house. The contractors were obviously required to complete the job as quickly as possible, before anyone could raise the alarm, and so nothing inside the house – not the furniture, the family records, or the quality kauri, matai, rimu, totara and jarrah timber – was salvaged.

The New Zealand Herald subsequently reported 'the demolition of one of the city's grand old landmark houses'.[13] Despite having the power to save the building, the Auckland City Council had acted in a 'spineless' manner[14] and done nothing. Its deference to market forces at the expense of heritage echoed the previous year's destruction of three Spanish mission-style villas in St Heliers, despite claims they were an essential part of the neighbourhood character.[15] Obviously, a new attitude was needed, one that would encourage owners to adapt, rather than destroy, buildings of heritage value.

A passionate appeal

Following the demise of Coolangatta, Auckland conservation architect Adam Wild suggested there needed to be a shift from 'an assumption of demolition to an assumption of conservation'.[17] In 2010 Coolangatta was the subject of a book by lawyer Peter Macky, a grandson of the house's original owners, Alfred and Jessie Foster. Both a tribute to the house and a history of those who had lived in it, it was also a passionate appeal for greater appreciation of Auckland's heritage.[18]

'A PLUME OF DUST'

Coolangatta was demolished in just 18 minutes, causing a plume of dust to rush down the valley. One neighbour recalled: 'It looked like we'd been flour-bombed.' Remuera heritage researcher Jennifer Hayman provided another perspective, describing the loss of the house as 'a sad indictment on the perceived short-term gain … in conflict with heritage and national identity'.[16]

Coolangatta has gone, but several of Noel Bamford's other notable buildings have survived. He was a busy man, one of the first lecturers at the School of Architecture which was established at what was then Auckland University College in 1918,[19] and a member of the board of Maori Arts and Crafts.[20] In 1932 he gave a public talk, 'How to Look at Pictures', at the Auckland Art Gallery, informing an appreciative audience that the only way to understand a picture was to 'surrender oneself' to its influence. He did not consider a knowledge of technique necessary, for 'a work of art should make an instant appeal', but he did concede that painters now tended to express 'unusual ideas' and only the initiated could follow their intentions.[21]

In 1919 Bamford had been member of a committee to consider a suitable form for a war memorial for Auckland, resulting in the decision to erect a new museum on what was then known as Observation Hill in the Domain.[22] The Auckland War Memorial Museum opened in 1929, and two years later Bamford commented on a proposal to floodlight the building. He approved of the idea of 'a stately pile of amber light in our wonderful white Pacific night', but wondered whether such dramatic effects might result in 'the delicate nocturne, the comfortable pace, the sun by day and the stars by night being relegated to the limbo of lost things'.

Ominously, Bamford seemed to anticipate the sort of forces that would result in the demolition of Coolangatta three-quarters of a century later: 'We are entering upon a period of industrial, moral and aesthetic anarchy, confounding more and more the good with the bad, the beautiful with the ugly, the true with the false, and knowing not what is what, find ourselves at the mercy of a commercially-minded world showering upon our bewildered heads the products of a self-created mass production.'[23]

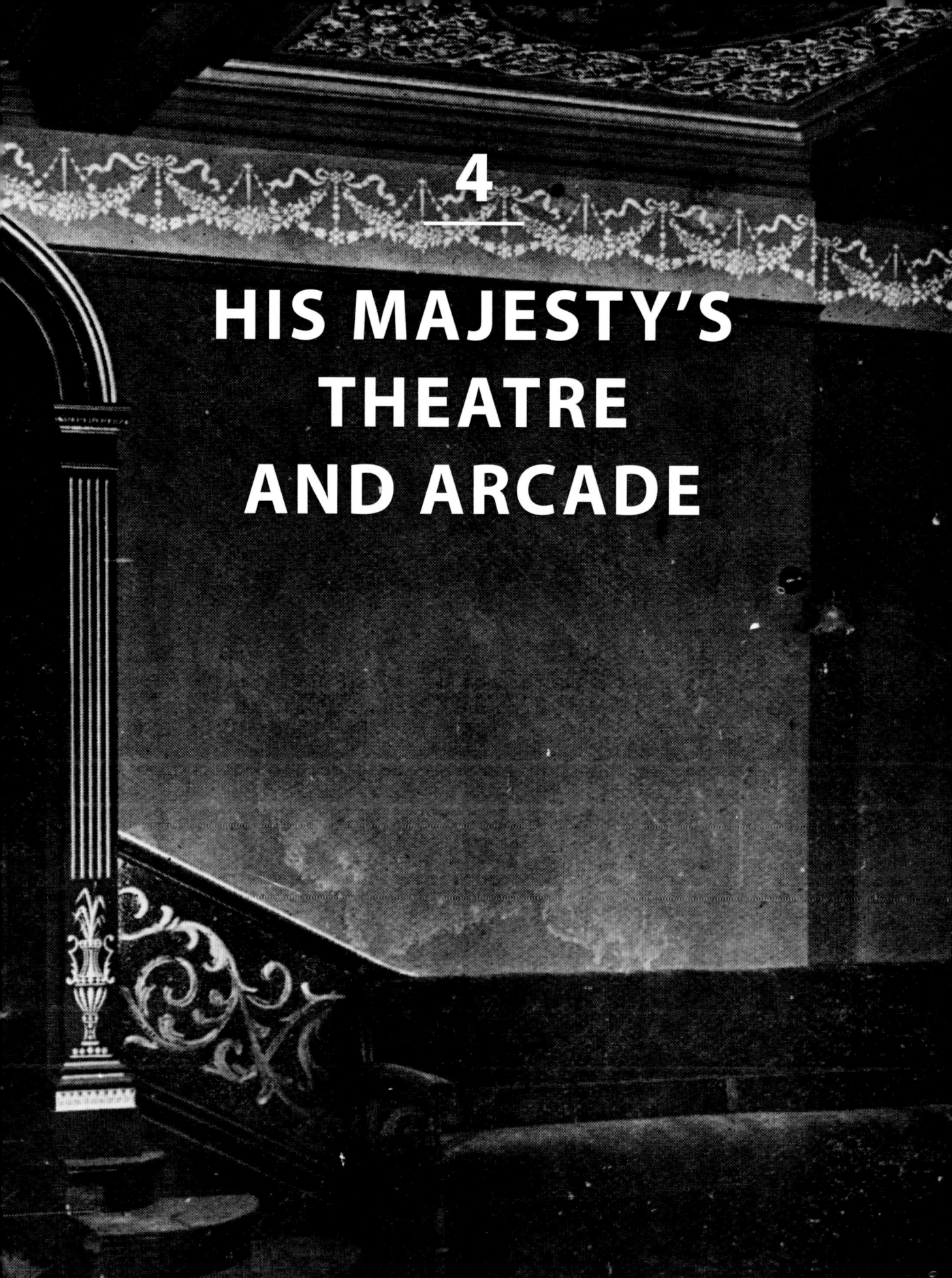

4

HIS MAJESTY'S THEATRE AND ARCADE

A FINE PILE OF BUILDINGS:
HIS MAJESTY'S THEATRE AND ARCADE

THE 1980s WAS A PERIOD OF EXCESS in Auckland, as its urban landscape was blighted by the arrival of high-rise mirror-glass towers. There was also an undue haste when it came to the removal of heritage buildings, and no more so than in the case of His Majesty's Theatre and Arcade. The theatre opened in late 1902, impressing with its sumptuous decorations and providing entertainment for Aucklanders for the next 84 years. But over time it had been allowed to fall into disrepair, and when demolition seemed increasingly likely in 1987 it became the focus of a heated debate.

Those offering support for the preservation of the theatre included actors who had trod its boards, and also Prince Charles. But His Majesty's was at the mercy of various forces – both active and inactive – involving owners, developers, the city council and public, and its fate was sealed when it was torn down over a weekend.

In late 1901 plans were announced for a new theatre and arcade for Auckland. His Majesty's, honouring Edward VII who had recently ascended the throne following the death of Queen Victoria, would be built on a three-quarter-acre section next to the Metropolitan Hotel in Queen Street. It would be designed by Australian architect William Pitt, with supervising architect Edward Mahoney & Son of Auckland, and cost £20,499.[1] Behind the project were three businessmen: Robert Henry Abbott, Francis Kneebone and a future mayor of Auckland, Arthur Mielziner Myers, who was managing director of the Campbell and Ehrenfried brewing and liquor business, and later the driving force behind other major developments in the city, including the new Town Hall and Grafton Bridge.[2]

A spacious arcade
Old shops in Queen Street and saleyards in Durham Street had to make way for His Majesty's. They were replaced by a spacious arcade, lined with shops and offices and lit with 48 incandescent lights. It ran back from Queen Street, leading to a flight of

ABOVE: *His Majesty's Arcade, looking from the Queen Street entrance towards the His Majesty's Theatre, c.1982.*

PREVIOUS: *A grand entrance to His Majesty's Theatre, showing elaborate classical detailing and decorated ceiling.*

EARLY QUEEN STREET

Auckland began in 1840 as a cluster of tents and small wooden buildings on the shore of Commercial Bay, and its initial development was to the east, along Shortland Crescent (later Street). Early Queen Street ran through a gully, following the course of the Waihorotiu Stream, which was later covered over. Before long the street became the main commercial centre, extending southwards towards the Karangahape Road ridge. In 1900 the first motorcar took to the street, and two years later it received its first electric tram service. By the late 19th century Auckland was booming, and had become New Zealand's largest industrial centre. This growth was reflected by such recent imposing buildings in lower Queen Street as the Victoria Arcade and the New Zealand Insurance Company head office (both demolished), which were joined in 1902 by His Majesty's Theatre and Arcade.

stairs which ascended to the dress circle and gallery of the theatre. William Pitt was a theatre specialist, and His Majesty's would be similar to – although smaller than – the Princess Theatre in Melbourne, which he also designed. Seating would be arranged so that a good view of the stage could be obtained from any part of the auditorium.[3]

The developers promised to spare no expense in ensuring their new theatre was 'commodious and convenient' and 'one of the handsomest structures of its kind'. Novel features included a sliding roof designed to eliminate stuffiness, and fibrous plaster ceilings which were not only decorative but fireproof and soundproof. It was also equipped with the latest in electric lighting; a gas-powered dynamo was capable of lighting 550 lamps, each of 16 candle power, while the auditorium was lit by 100 lamps in the dome, and the stage was illuminated by four borders of footlights.[4]

In terms of size and appearance, His Majesty's Arcade was intended as 'an exact counterpart' of the recently completed Strand Arcade, further up and on the same side of Queen Street. The new theatre would be leased, and there was a strong demand by tenants for the shops, offices and clubrooms in what the directors termed 'this fine pile of buildings'. An early occupant was the recently formed Commercial Travellers' and Warehousemen's Association, while other organisations that followed were the Auckland Chess Club and the Amateur Sports Club.[5] And just as the earlier Victoria Arcade, on the opposite side of Queen Street, had attracted artists, one of His Majesty's early tenants was Miss Ellen von Meyern. She opened a studio where she took pupils and exhibited her own watercolour portraits of Maori.[6]

Height and airiness

A visitor to His Majesty's Theatre prior to the official opening found its decoration 'very gorgeous and florid' and obviously expensive – the large gilt proscenium in particular. The theatre impressed with its height and airiness, and the sense of size was emphasised by the climb up to the amphitheatre and its many 'gods'. Such features prompted the

View of the stage and proscenium from the dress circle at His Majesty's Theatre shortly before its completion in late December 1902.

claim this might be the finest theatre south of the line and, further, that there was nothing comparable in any English city of a similar size to Auckland.[7]

The J.C. Williamson Musical Comedy Company provided the opening show for the new theatre. *A Runaway Girl*, which had completed a two-year run in London and also played in Australia, Wellington, Christchurch and Dunedin, was brought to Auckland by a company of some 80 English and Australian operatic artists, supported by an opera chorus, orchestra and ballet. It provided opportunities for 'sumptuous staging' and was said to be distinguished by the quality of its musical items. One of these was 'The Soldiers in the Park', a swinging march song that accompanied British troops when they paraded up Queen Street during a visit to Auckland in February 1901.[8]

A full auditorium at His Majesty's Theatre for the first night of A Moorish Maid, *on 26 June 1905. The romantic opera, by New Zealander Alfred Hill, ran for a season of six nights and its cast included an Arabian chorus and Nubian slaves.*

A palpable hum

His Majesty's opened with much anticipation on Boxing Night 1902, and was filled by what was claimed to be the largest audience ever gathered in a New Zealand theatre. There was a palpable hum as patrons arrived and admired the 'luxurious appointments, lovely mural decorations and frescoed roof, and richly gilt proscenium'. The three directors of the company, along with the general manager of the National Bank, sat in a box to the right of the stage. There was less evening dress among the audience than might have been expected on opening night, perhaps due to another attraction, a local race meeting.[9]

The *Auckland Star* reviewer found the theatre 'far more handsome' than could have been imagined, and the removable roof slid off during the interval, providing welcome ventilation for the audience. But one technical innovation that didn't go as planned was the electric light system, plunging the auditorium into darkness for a period. Management reassured the audience that there was no danger, and then resorted to the older technology of gas lighting.[10]

The new theatre was impressive, but the same could not be said for its opening show. Although described as 'gorgeous, very radiant, very tuneful', *A Runaway Girl* was also judged 'extremely absurd' and crowded with characters who had little to do with its 'shred of a plot'. In fact, its 'funnyisms' were accused of being 'not sufficient to justify their existence', while the 'inanities of the dialogue pall dreadfully after three hours'. Many of the jokes – involving ladies' underwear, split pants and babies, 'inopportune and otherwise' – were said to be half a century old, with one reviewer admitting to a personal preference for 'open indecency' rather than double entendres.[11]

The Williamson organisation may have thought it had a winner in *A Runaway Girl*, and early full houses suggested it was set for a good season. But it received only a 'very fair attendance' on 1 January, by which time public curiosity to see the new theatre may have outweighed enthusiasm for the play itself.[12] His Majesty's Theatre was also in direct competition with the Opera House and a music hall, and immediately after the Christmas and New Year period there was a notable shrinkage in the audiences at all three venues. *A Runaway Girl* survived for 10 days, which was considered a respectable run in a city with a population of 60,000.[13] The final performance was on 6 January, and the following night it was replaced by a new show, the Chinese musical comedy *San Toy*, whose cast of characters included the delightfully named Sir Bingo Preston, British Consul at Pynka Pong. Nothing if not versatile, members of the cast of *A Runaway Girl* quickly adjusted to their new roles in *San Toy*, hailed 'The Most Popular and Successful Comic Opera of all Time'.[14]

A wide range of performances

An early test for His Majesty's was the collision of a cart pulled by a runaway horse with one of the verandah posts supporting the Arcade.[15] But the theatre became established, bringing a wide range of performances to Auckland. The audiences also varied, prompting one theatre-goer ('Disgusted') in 1907 to complain about young men who had brought in a stock of beer to consume during the performance.[16] On the other hand, His Excellency the Governor (Lord Islington) attended the Auckland Amateur Operatic Society's performance of *The Pirates of Penzance* at His Majesty's in 1911, while a subsequent governor-general, the Earl of Liverpool, and the Countess of Liverpool, experienced the pantomime *Dick Whittington* there in 1918.[17]

By the end of the 1920s the company operating His Majesty's Theatre and Arcade was in a healthy position, with property valued at £53,000. Revenue continued to rise, although by the early 1940s net profit was reduced by taxation and compulsory war damage insurance.[18] There was a continuing demand for rented properties in the Arcade, but the theatrical business was now facing increasing competition from the 'talkies'. Ownership of His Majesty's Theatre passed to the estate of Robert Abbott, one of the original developers, and in 1941 it was sold to J.C. Williamson Ltd. There were plans to give the theatre a facelift, and one certainly happened in preparation for the presentation of *My Fair Lady* in 1961, when much of the decorative work, such as the proscenium arches and plaster decoration, was removed or repainted. At some point the sliding roof was covered over, presumably because it leaked, and the richly painted canvas which once covered the dome was removed.[19] J.C. Williamson had expected an Auckland season of three to four months for *My Fair Lady*, but it lasted six months, and when the curtain came down on the final performance it had been seen by 640,000 people.[20]

Obvious signs of neglect

In the late 1970s the J.C. Williamson organisation was wound up. In preparation for selling the property, it managed to get the New Zealand Historic Places Trust rating of His Majesty's reduced from B to C, which meant the building had some significance but not enough to warrant protection.[21] In 1981 the theatre was sold to Kerridge Odeon, the New Zealand cinema chain established by (Sir) Robert Kerridge and H.B. Williams in 1926. The building was now deteriorating and showing obvious signs of neglect; the high vaulted glass in the Arcade had been replaced by now-discoloured plastic sheets, and the tiling underfoot had long been asphalted over.[22]

In 1987 the Kerridge and Williams family interests in the company were acquired by a new company, Pacer Kerridge. Rumours of the fate of His Majesty's were now rife, and it was announced that it would be auctioned the following February. The removal of fittings from the building heightened suspicions that demolition was imminent, and it seems a permit to do so had been applied for from the Auckland City Council on the afternoon of 23 December, immediately prior to the Christmas holiday period. Concerned citizens kept an eye on developments, and around

> **AN EX-PRIME MINISTER ON THE STAGE**
>
> Of the many entertainers who appeared on the stage at His Majesty's during its 84 years, one of the most unlikely was an ex-prime minister. In July 1986 Robert Muldoon appeared in 10 performances in front of 13,000 people as narrator in *The Rocky Horror Picture Show*, one of the final shows at the theatre. As Barry Gustafson has recorded, Muldoon began somewhat hesitantly and lacked in confidence, but went on to receive 'thunderous ovations' for his attempts at a pelvic thrust.[23]

The first swing of the wrecker's ball and the demolition of His Majesty's Theatre and Arcade is under way, in January 1988.

midday the Historic Places Trust served a notice on Pacer Kerridge that the building was proposed for a B classification. Strictly speaking, demolition could not legally be carried out until at least 5 January, when the city council permits office reopened and could consider the matter.[24]

Overseas support for His Majesty's came from British actor Windsor Davies, who had performed in the theatre earlier in 1987 in *Run For Your Wife*, and was reportedly putting together a telegram from other British entertainers.[25] Her Majesty's eldest son also expressed support for the cause, and in a letter to the Civic Trust by his deputy private secretary, Prince Charles described the building as 'an important asset of the city's heritage' and hoped the Auckland City Council would give 'due and careful weight to preservationist arguments'.[26]

'A rat-infested dump'

An *Auckland Star* editorial noted that the Auckland City Council town planning committee had decided His Majesty's was not worth preserving, and had done so without consulting the Historic Places Trust, the national body established by statute to make decisions on such matters.[27] Five days later the same newspaper published contrasting views on the matter of preservation. Theatrical entrepreneur and Auckland City Councillor Philip Warren summed up the building as 'A rat-infested dump; aesthetically of no value; it leaks; sight-lines are appalling; electrically hazardous – how it had avoided enforced demolition orders of the past 10 years is beyond me.' Warren had presented over 20 productions in the theatre, from Barry Humphreys to *Snow White and the Seven Dwarfs*, and claimed not one of these artists had suggested to him the place was worth saving.

Warren pointed out that the theatre no longer resembled the original building of 1902; in addition to the alterations for *My Fair Lady*, it was 'rebuilt' for the tribal love-rock musical *Hair* in 1972. Warren had been involved in the saving of several theatres throughout New Zealand, including the Theatre Royal (Christchurch), the State Opera House (Wellington), and the Regent (Dunedin), and claimed all these – and others – were 'far superior to His Majesty's'. He suggested heritage organisations should turn their attention instead to Auckland's St James and the Civic Theatre.[28]

Stuart Mackenzie, representing the New Zealand Entertainment Artistes' Benevolent Fund, countered by pointing out that His Majesty's was the only facility in Auckland suitable for a very large range of shows. If demolished, the city would face 'a theatrical disaster', losing both local productions and overseas touring companies. Mackenzie also suggested that the departure of the building would be to the detriment of Queen Street, which was 'rapidly becoming an abomination'.[29] The *Auckland Star* agreed with the latter sentiment, observing that Aucklanders had allowed their city centre to become 'soulless glass and steel', and would probably accept the inevitability of His Majesty's being replaced by another mirror-glass building.[30]

Demolition permit issued

There was the suspicion that the Auckland City Council favoured the removal of His Majesty's in order to reduce competition for the Aotea Centre for the performing arts, then under construction. It was not surprising then, that, the council issued a permit for the demolition of the theatre and approval for the closure of adjacent streets for the New Year holiday break. Recognising it had no chance of success, the Historic Places Trust now decided against any taking legal action to prevent demolition proceeding.[31]

Neither of Auckland's two main newspapers considered His Majesty's worth saving. *The New Zealand Herald* suggested Prince Charles had been 'incautious' in lending his name to the campaign to save the theatre, and had not been well briefed on the subject.

The prince claimed the theatre was among the finest in Auckland, but the paper pointed out the Arcade may well have been its 'seediest'.[32]

Meanwhile, the *Auckland Star* blamed the present situation on the city council and the apathy of its citizens. While it complained that His Majesty's seating was uncomfortable and views of the stage from many seats were obstructed by pillars, the fact that it was booked for shows for 85 per cent of 1988 suggested that it was in demand. The paper therefore suggested that a condition of any new development on the site should be that it contain a new live theatre with the same seating capacity which, it pointed out, would not compete with the Aotea Centre, which was not designed as a theatre.[33]

The *Auckland Star* believed Aucklanders had allowed buildings 'of far greater aesthetic and architectural worth' than His Majesty's to be torn down, and was astonished to see such controversy over a 'tired, dilapidated, poorly designed, uncomfortable theatre'. However, it could not condone what appeared to be an attempt to begin demolition in the dead of night without the required permit. That act brought no credit to Pacer Kerridge, who presumably intended to pre-empt any action by the Historic Places Trust or the city council that might affect its plans to auction the site.[34]

The wreckers begin

Demolition would have begun on 3 January 1988 but was thwarted by protestors on the roof of the theatre. However, it began in earnest two days later, and it took several swings of the wreckers' ball to produce a crack in the back wall of the building – which, ironically, previous owners had claimed was an earthquake risk. The theatre was reduced to rubble on the weekend of 10–11 January.[35]

To add insult to injury, the site once occupied by His Majesty's Theatre and Arcade lay vacant for many years until occupied by City Life Apartments. There may have been some sense of justice in 1992 when the company responsible for the demolition of the theatre, Pacer Kerridge, went into receivership. On a more positive note, the publicity surrounding the controversial demolition did – but not for the last time – highlight the issue of the preservation of historic buildings.[36]

Among other theatres by architect William Pitt, his Napier Municipal Theatre (1912) was destroyed in the 1931 earthquake, but his Wellington Opera House (opened in 1914) survives.

5

KILBRYDE

THE MOST BEAUTIFUL VIEW:
KILBRYDE

THE FIRST HOUSE occupied by Sir John Logan Campbell, the 'Father of Auckland', has been preserved in Cornwall Park, which he gifted to the people of New Zealand in 1901. But Campbell's last and most impressive house only managed to outlive its owner by a dozen years. In June 1924 this magnificent Italianate residence was hauled down from its cliff-top location overlooking the harbour to provide spoil for a rapidly expanding city. The public outcry in Auckland that accompanied its destruction foreshadowed future controversies surrounding the demolition of outstanding buildings such as His Majesty's Theatre and Arcade and Coolangatta.

John Logan Campbell was born in Edinburgh, Scotland, in 1817, and, following his father's footsteps, studied medicine at Edinburgh University. But instead of practising in his homeland, in July 1839 he sailed as a surgeon on the emigrant ship *Palmyra* to Australia. From there he crossed to New Zealand, and at Coromandel met up with fellow Scot and future business partner, his *Palmyra* shipmate William Brown. In mid-1840, with two companions and a Maori guide, the pair sailed west into an inlet off the Hauraki Gulf.

Half an hour later the party entered a deep bay, which they were told was Orakei, and landed on a small shelly beach at the base of 'some lovely wooded slopes'. Those slopes would soon be absorbed by a growing settlement, and Campbell would become its most prominent citizen. Half a century after first stepping ashore at Orakei he would be acclaimed 'Father of Auckland'.

Auckland's first merchant business
The canny Campbell and Brown anticipated Auckland would become the capital of the colony, and on 21 December 1840 they established its first merchant business, in Commercial Bay at the foot of present-day Shortland Street. Campbell's first home in the settlement was a tent, from which he graduated to a raupo whare. In O'Connell Street the pair built Acacia Cottage, where Campbell lived for a period with Brown and his wife.

ABOVE: *John Logan Campbell outside his home Kilbryde, on Campbell's Point, Parnell, in October 1905.*
PREVIOUS: *The Italianate splendour of Kilbryde, overlooking the Waitemata Harbour with North Head and Rangitoto on the right, in 1885.*

THE LOVELY EXPANSE OF WATER

John Logan Campbell recorded his first impressions as he entered Auckland harbour in mid-1840: 'Ah! never can I forget that morning when first I gazed on the Waitemata's waters. The lovely expanse of water, with its gorgeous colouring, stretched away to the base of the Rangitoto, whose twin peaks, cutting clearly into the deep blue sky, sloped in graceful outline to the shore a thousand feet below. Still further distant we saw a bold round high headland, backed by a still higher hill, and far away before us a long expanse of glancing waters as far as the eye could reach … We rowed up the beautiful harbour, close in shore. No sign of human life that morning; the shrill cry of the curlew on the beach and the rich full carol of the tui or parson-bird from the brushwood skirting the shore fell faintly upon the ear. The sea was smooth as glass, and the flood-tide swept us along.'[1]

In 1920 the cottage was relocated to Cornwall Park where it survives as the oldest timber dwelling in Auckland and one of the earliest in New Zealand.[2]

In 1841 Campbell bought a cliff-side section in Jermyn Street (later to be Anzac Avenue), overlooking Mechanics Bay, and built a four-roomed cottage named Logan Bank. From 1856 to 1871 he spent most of his time in Europe, mainly Italy, and shortly before his return – now with a wife, Emma Wilson – he arranged for additions to be made to

Logan Bank. These qualified as Auckland's first structure to be made of poured concrete, a material Campbell had seen in England and on display at the 1870 Paris Exhibition.[3]

The Campbells lived at Logan Bank for some 11 years, before moving a little over a kilometre due east to another cliff-top location. Campbell has been described as 'a competent amateur architect', and in 1881 he sent plans for his next and final residence to the Auckland architectural firm of Edward Mahoney & Son, where they were revised and working drawings prepared.[4]

A likely inspiration for the new house was The Pah, at Hillsborough, built in 1877 for businessman and chairman of the Bank of New Zealand, James Williamson, and designed by Thomas Mahoney, son of Edward Mahoney. It was Italianate in form, with an Ionic entrance portico and a tower. In the words of John Stacpoole, 'other magnates liked what they saw', and Campbell's new house would also be Italianate and distinguished by a tall tower. And while considered less successful externally than The Pah, its remarkable features included a fresco room, which had all its walls painted with landscapes, and a minstrel's gallery.[5] That Campbell's house was Italianate in style was hardly surprising given the time he had spent in that country, while the view from it of the symmetrical cone of Rangitoto against the Waitemata was said to be reminiscent of the coast of Italy.[6]

One of Auckland's finest views

In 1881 the Campbells moved in. Their new house, at the foot of Gladstone Road, was named Kilbryde, after a family castle in Perthshire, Scotland. The site, which became known as Campbell's Point, was initially an exposed promontory covered with tea-tree and low scrub. It is said that Campbell personally cut down the scrub, planted trees and formed paths around the cliff edge. A particular attraction for him was the view to the east of the small bay where he had first landed in Auckland, in 1840, and also of Motukorea (known as Brown's Island, having been purchased by Campbell's business partner in 1840). To the west Campbell was able to watch the steady growth of Auckland and the infilling of the harbour, a process that would eventually absorb much of his own property.[7] The view was considered one of Auckland's finest, as would have been attested by those fortunate enough to be invited to Kilbryde in August 1908 to watch the arrival of a fleet of American warships in the Waitemata Harbour.[8]

John Logan Campbell lived at Kilbryde for about 31 years, dying there in his sleep on 22 June 1912. He departed a little before dawn, when the waters of the Waitemata were 'just at the last of the ebb … The tide and his dauntless spirit went out together.' Less than two months later Lady Campbell died, also at Kilbryde, the result of a fall shortly before her husband's death.[9]

Later the mayor of Auckland, Sir James Parr, suggested Kilbryde as a summer residence for the governor. He believed His Excellency's presence during that period gave the city

The elegantly appointed drawing room at Kilbryde.

a certain prestige and brightened its social life, while also being good for business. Parr considered the present Government House property too valuable to be used for only three months in the year and suggested, presciently, that it could be put to a more useful purpose such as part of the University of Auckland. He considered Kilbryde the ideal alternative, and in fact 'in the whole of Australasia he did not know of a more beautiful spot'.[10]

The harbour board steps in

But Kilbryde did not become a vice-regal residence. It was not included among Sir John's many public benefactions, and was bought from the trustees of his estate by the Auckland Harbour Board.[11] The board hoped to be able to exchange it with a portion of Campbell's Point already secured by the city council as a public reserve, so it could demolish the headland and use it for harbour reclamation. There were strong objections to the scheme, the *Auckland Star* noting: 'Of all the headlands by which the coast of the Waitemata is so charmingly broken and diversified, none can compete with Campbell's Point. The view from the Point by day or night is one of unsurpassed loveliness.'[12]

But the Point remained under threat, and now from a proposed new railway outlet for the city. After negotiations with the harbour board it was decided that the Railway Department would take the whole of Campbell's Point, comprising parts of the public reserve, Gladstone Road and the Kilbryde Estate, while the Auckland City Council would purchase the balance of the latter, which included the house, for £5250.[13] The city council later combined it with an adjacent property, purchased under the Public Works Act, to make a public park.[14]

Much public opposition

Again, there was much public opposition to the proposal to remove Campbell's Point, and citizens made their feelings known by signing a petition at the Victoria Arcade.[15] But by March 1915 demolition was well under way, resulting in what was described as 'an ugly gaping wound …as… the harbour's beauty spots are being demolished to make way for the iron rails'.[16] The Point was disappearing at the rate of some 200 tonnes per day. Gelignite was used to topple the sandstone cliffs, which were then broken up by men wielding picks and crowbars. Before long, all that remained was a steep bank, 21 m high, which was about 30 m from the old Kilbryde residence.[17]

And while the fate of that house lay in the balance, Logan Campbell's previous residence was also in the news, following plans to convert Jermyn Street into another outlet for the city – for road traffic. New Zealand was now at war, and because of its Germanic reference the name of the street also had to go. Suggested alternatives included Campbell Road, in recognition of the previous owner of Logan Bank, and Britomart Street (or Road), in remembrance of the fort that once had stood on the Point of the same name (demolished in the 1880s) beyond present-day Emily Place.[18] In the end the new outlet was given the patriotic name of Anzac Avenue.

On 6 November 1918 the deadly worldwide influenza epidemic 'laid its clammy hands' on Auckland. The hospital board accepted the city council's offer of the use of Kilbryde as a temporary hospital, and it was quickly converted from a 'large, dusty, vacant house' to 'a well-ordered hospital'.[19] By the end of November 1918, when there were signs that the epidemic was being 'throttled slowly', staff had attended 121 cases at Kilbryde, of whom 52 had died and 35 were still being treated.[20] By early December the house was empty, but a year later there were suggestions that the hospital board retain it in case of another similar emergency. In fact, in February 1920, following concern that influenza was on the rise again, the old house was disinfected and prepared to accommodate 60 patients, just in case.[21]

Kilbryde is demolished

By now Kilbryde was in need of extensive repairs and repainting, and the Reserves Committee of the city council adopted the recommendation that it be demolished. At the same time it was suggested that a drinking fountain, or some other suitable memorial be erected to mark the place where Sir John Logan Campbell had lived.[22] But Kilbryde wasn't finished, yet. In 1921 the Education Board investigated its use as a hostel for women students at the Auckland Teachers' Training College, but did not proceed with the idea on account of the cost of refurbishment.[23]

Three years later, in order to make a more effective public park, the city council reaffirmed its decision to remove Kilbryde, along with certain trees that were obstructing harbour views. However, the plan was to retain Sir John's favourite tree, a large pohutukawa which stood on the lawn beside the house. And while admitting that it might 'more or less' entail the destruction of what remained of Campbell's Point, it was suggested that the making of a gentle slope from the harbour side of the property would compensate for the loss and, further, would have received 'prompt support' from Sir John himself.[24]

With Kilbryde's fate now sealed, the Parks Committee proposed that materials salvaged from its demolition be used in the construction of much-needed houses for workers. In response, Michael Joseph Savage, a member of both the Auckland City Council and hospital board and Member of Parliament for Labour, advised that 'not a worker in Auckland' wanted Kilbryde to be dismantled. The council was also informed that homeless persons had now taken possession of Kilbryde, and if it was demolished there would be 'a mild revolution' in the city.[25]

Kilbryde was finally demolished in June 1924.[26] That event, compounded by the earlier destruction of Campbell's Point, was too much for some citizens. The *Auckland Star* claimed the harbour had been 'changed by the magic hand of the city council into a veritable slum. Where is the magnificent outstanding bluff, Point Campbell? Where is the fine expanse of sea between it and the North Shore? Where is Kilbryde?'[27] Four decades later another writer lamented Auckland's lost buildings, among them 'the grandest of all houses', the 'elegant white mansion' that was Kilbryde. That same writer recalled Sunday School picnics on the lawns of the house, and children peering in through the windows at the lofty ballroom, the great hall and the servants' quarters.[28] Today Kilbryde is long gone, but a remnant of its once expansive grounds remains as part of the Parnell Rose Gardens.

After the Campbells had moved to Kilbryde, their previous home, Logan Bank, served as a boarding house.[29] In 1917 it was auctioned for removal, after which it was dismantled and its timber, sashes, doors and windows were sold on site.[30] Logan Bank's pioneering concrete structure appears to have fallen into disrepair, and the property was later bought by another former mayor of Auckland, Sir Ernest Davis. In 1960 he presented it to the city council with a view to it becoming a public space called the Ernest Davis Lookout. A year later, the remains of Logan Bank were described as a 'battered ruin, festooned with creepers and surrounded by thick undergrowth'.[31] Today the Lookout, at 110–116 Anzac Avenue, is dominated by a steep path that zig-zags down to Beach Road. From this elevated property Campbell enjoyed a panoramic vista of the Waitemata, a view now largely blocked by buildings as Auckland continues to develop.

6
PARTINGTON'S MILL

SENTINEL OVER AUCKLAND:
PARTINGTON'S MILL

'BLOW HIGH, BLOW LOW, the sails go merrily round, never tiring, taking their motive power from the Great Almighty, and preparing the staff of life for his Creatures.'
– *Observer*, 4 March 1882

The best known of all Auckland's – if not New Zealand's – needlessly demolished buildings is a dubious honour claimed by Partington's Mill. In addition to being a functioning mill, it served as a distinctive landmark on the ridge above the Queen City for almost a century, until it was controversially hauled down in 1950. Although it altered the cityscape, that act of vandalism did at least stimulate a wider interest in the nation's heritage. But the mill itself has not been forgotten; there were plans to erect a replica in 1978. And whenever another significant old Auckland building bites the dust, reference is invariably made to three earlier losses: the Victoria Arcade, His Majesty's Theatre and – heading the list – Partington's Mill.

The origins of Auckland's best known and most durable landmark lay with the arrival of Charles Frederick Partington, from Oxfordshire. In 1847 he went into business with John Bycroft, taking over Auckland's first flour mill, in Mt Eden, in what became Windmill Road. The partnership with Bycroft (who later became a national name in biscuits) lasted until late 1849, and Partington carried on the business until May 1850. He then bought land in Symonds Street and, with borrowed money, built a large, well-equipped windmill.

Good sculptural form
The mill was constructed of bricks made from clay dug on the site and built by Henry White, who had arrived in Auckland from Cornwall in April 1843.[1] White was also responsible for a number of other notable buildings in Auckland, some of which have survived (Carrington Psychiatric Hospital and Pitt Street Methodist Church) and some which haven't (St Paul's Church).[2] The identity of the designer of the mill isn't known, but because of its resemblance to such structures in Suffolk, architectural historian John

ABOVE: *Partington's Mill showing lifting machinery in position and the first stage of demolition, the lowering of the fantail (the device which orientated the sails to the wind), in April 1950.* PREVIOUS: *Looking south up Liverpool Street towards the Karangahape Road ridge, dominated by Partington's Mill.*

Stacpoole has suggested that William Mason (who was responsible for the Eden mill) was involved. Stacpoole has described Partington's Mill as 'the perfect example of good sculptural form derived from purely functional requirements', some half-century before modern architects and industrial designers were espousing such concepts.[3]

The new mill was in operation by August 1851, when flour (in various grades: fine, seconds and sharps) as well as meal, crushed maize and bran could be bought at the 'New Brick Windmill'.[4] The business grew, and in 1856 it became the grandly named Victoria Flour Mills and Steam Biscuit Factory, while the retail side was covered when Charles's brother Henry opened a shop in Queen Street. In order to be free of the vagaries of the wind, and both for reasons of economy and with a superior product in mind, Partington equipped his mill with the latest steam-powered machinery, imported from Reading ('famous for biscuit') in England.[5] By the mid-1860s the windmill was a familiar sight to locals, and 'eagerly looked for from the deck of many a coasting craft nearing the North Head'.

With the onset of hostilities between Maori and Pakeha in the Waikato in the early 1860s, the supply of locally grown wheat ceased. The mill's sails may have stopped turning, but a local paper advised that it had not become a 'castle of indolence'. Partington was now concentrating on baking biscuits in what was said to be the biggest such business in Auckland, his customers including the Imperial and Colonial troops. His steam-powered machines mixed 150 kg of flour in 10 minutes, after which the dough was flattened into long sheets and cut into biscuits at the rate of 350 in five minutes. A staff of two men and three boys controlled the operation, and could produce two tonnes of biscuits in just 10 hours. And just in case some wheat did come available, the steam engine could also power the flour mill.[7]

By 1866 a bag of wheat could be ground and baked into a biscuit in less than two hours. Partington's business was described as being 'nestled in umbrageous shrubbery and trees',

> **'THE OUTSKIRTS OF TOWN'**
>
> At that time Auckland barely extended beyond Shortland Crescent (now Shortland Street) and the western side of Queen Street, and where the mill stood was a 'distant suburb, a waste of ti-tree scrub'.[6] Early Auckland was centred on the shoreline of Commercial Bay, at the bottom of Queen Street, and Partington's Mill was built on what were then the outskirts of town. It stood near the intersection of two main roads, Symonds Street and Karangahape Road. To the east it overlooked the Symonds Street Cemetery, established in 1842 and the first official cemetery in Auckland, and Grafton Gully, which was first spanned by a (pedestrian) bridge in 1884. To the west, Partington's Mill had a bird's-eye view along Karangahape Road, which can claim to be the oldest street in Auckland. Long prior to European settlement, this ridge provided Maori with a route between the region's two harbours, the Waitemata and Manukau.

and almost hidden from the view of passers-by. The prominence of the windmill was highlighted in May 1869 during celebrations marking Queen Victoria's 50th birthday. Chinese lanterns and coloured lamps were displayed throughout Auckland, and one of the more successful illuminations was the mill, where lamps in the windows gave it the appearance of 'a huge lighthouse'.[8]

The mill in disrepair

His biscuit business was booming, and in 1869 Charles Partington, along with two of his sons, was tempted by opportunities on the Thames goldfields. But when the venture did not work out he decided that his mill – which had fallen into disrepair – should be sold off and the land subdivided into building allotments.[9] There was strong public reaction to the prospect that the building that had 'kept sentinel over Auckland for so many years' seemed destined for destruction. Partington was accused of possessing 'an obdurate and a hard heart' and having no 'refined sentiment'. At the time it was noted that one of the arms of the mill had become detached and was dangerous, and had reduced property values in the vicinity.

Among those who would miss the old mill were captains of vessels who had been away from Auckland harbour for a time; they would be 'under some sort of hazy impression that they have lost their bearings, or have not kept their weather-eye open, which will cause them to swear sheets and halliards'. The disappearance of the mill would also inconvenience young gentlemen who depended on it to guide them on their erratic course home after a night out. Such was the feeling about the prospect of demolition that *The New Zealand Herald* suggested that the island of Rangitoto – 'Auckland's second great landmark' – could have been 'better spared than the old mill'. The same writer promised that if Partington were to retain the mill, his name would 'handed down to posterity in the eternal columns' in the same newspaper.[10]

The property was put up for auction in March 1873, but withdrawn when it failed to meet the reserve.[11] Two years later it was back on the market, for £650, which included Partington's 'handsome' eight-room house which fronted on to Symonds Street, and two adjoining allotments. Also, and 'to be sold cheap', were two pairs of French Burr mill stones, along with 'the necessary machinery for a complete mill'.[12] The machinery and surrounding land may have been sold off, but the mill was still available, and on the market again 20 months later. On this occasion it was advertised as containing about 90,000 bricks, in good condition, which would have to be removed by the purchaser unless he also bought the allotment on which the mill stood.[13]

Partington appears to have had a change of heart, and decided not to sell the windmill. But matters relating to its ownership became complicated when he died on 28 January 1877.[14] Determined to carry on the business, his sons Charles and Edward

Partington's Mill glimpsed through a vacant section between the Tivoli Picture Theatre (left) and the Maple Furnishing store in Upper Symonds Street.

raised capital and had the mill back in working order later that year.¹⁵ By 1882 the steam mill was reported to be able to handle spices as well as grain, while another machine dealt to wattle-bark, used in the tanning business. Wind-power was still used, although supplemented by a steam engine when necessary, while a strong wind was capable of producing about 25 horsepower, enough to operate all the mills.¹⁶

Meanwhile, Charles and Edward Partington had transferred their business interests to a mill belonging to the Auckland City Council at Western Springs, but this proved unsuccessful and by 1887 they were insolvent.¹⁷ A third brother, Joseph, had taken over the steam mill business back at Symonds Street.¹⁸ But when the Bank of New Zealand collapsed, the mortgage was foreclosed and the whole property sold to James Wilkinson, an engineer by trade and also chapel-keeper at the Wesleyan Church. Joseph Partington now rented from Wilkinson, but when he was reduced to a weekly tenant he moved his machinery from the windmill into the old biscuit factory, which he had on a more secure lease. He then continued to run his business by steam while Wilkinson attempted, unsuccessfully, to use the mill to polish heel- and toe-plates for boots and shoes.

Partington's grievances

Partington now wrote a letter to *The New Zealand Herald* asserting that Wilkinson had erected a building adjacent to the mill that was in contravention of local bylaws. Wilkinson countered, his letter allegedly 'bristling with lies, half-lies and deliberate misrepresentations', and indirectly accused Joseph Partington of dishonesty.[19] Unable to convince the editor of the *Herald* to give him the right of reply, Partington now commissioned a journalist, George Everard Bentley, to compile a pamphlet which outlined his grievances. This publication detailed how Partington had suffered at the hands of Wilkinson, who was 'a thoroughly treacherous landlord'. The latter had sold off land, effectively ruining the Old Mill by removing most of the surrounding space needed for necessary operations. The public had also been inconvenienced; in earlier times they had been able to use a private road as a thoroughfare from Symonds Street to Liverpool Street, but Wilkinson had now fenced it off. And whereas Partington had allowed photographers – amateur and professional – to climb the structure to obtain panoramic views of the City, Wilkinson no longer permitted them to do so.[20]

Following the appearance of the pamphlet, *The Story of the Old Windmill*, Partington and Bentley went on trial in the Supreme Court for publishing an alleged criminal libel. In summing up, His Honour suggested 'A more disgraceful or scurrilous thing never was written', but after a half-hour's deliberation, the jury found Partington not guilty. Bentley changed his plea to 'guilty', but because he had been unable to find bail and had already spent 12 weeks behind bars, he did not receive any further sentence.[21]

Back in control

Despite various obstacles, by 1910 Joseph Partington was back in control of the windmill, which was now in a state of delapidation. On a trip to England Partington bought machinery and parts for the mill, which was working again by mid-1916. He replaced the sails, and also added some 4.5 m to the height of the tower to provide better access to available wind, which had been reduced by the new buildings in the vicinity.[22] While a windmill might be considered 'quaint and out of place' in the centre of a city, its 'time-honoured' sails were said to connect the old Auckland with the new. And even when the sails weren't actually turning, the building continued to serve a very practical purpose, as demonstrated by a newspaper advertisement for lost property, mislaid somewhere between 'the top of Wakefield Street and the Wind-mill'.[23]

The matter of its ownership may have been settled, but the windmill continued to be plagued by problems. A strong gale in 1924 resulted in two of its sails having to be dismantled, and seven years later the mill was gutted by a fire that had spread from a neighbouring building. But the business was able to carry on, using a previously installed electric power plant, and the building was later repaired and the sails replaced.[24]

As well as being a local landmark, the windmill was a popular subject for local artists and students from the nearby Elam School of Art, while it also inspired poets. In 1928 'Ruthyn' reflected on the mill as a rendezvous for courting couples back in the 1850s:

> And the windmill sails were turning,
> As still they turn today.
> It watches Youth pass into Age,
> And sees Age pass away.[25]

Joseph Partington died in 1941, and without any apparent will. Just as when his father had died 64 years earlier, the ownership of the mill became complicated, and it was now part of an estate to be sold on behalf of beneficiaries. The old structure was at risk yet again, leading to the formation of the Old Windmill Preservation Society. Its committee included such leading citizens as E. Earle Vaile (chairman), F.N. Ambler, Dr Gilbert Archey (Director, Auckland Museum), John Barr (City Librarian and Curator of the Old Colonists' Museum), H. Goldie, and Professor C.R. Knight (Dean of the Faculty of Architecture, Auckland University College). Its publicity referred to Auckland's 'fast disappearing past'. It noted: 'New Zealand is a young country and because of its youth, has few monuments of an historical or sentimental value and Auckland has one in "The Old Windmill" in Symonds Street. It is a definite link with the early pioneers and is a relic of sentimental value for many …' In John Barr's opinion, 'It would be an irresponsible historical loss to Auckland if the mill was destroyed.' The society sought donations, which could be sent to the secretary, at the Town Hall, while supporting resolutions were passed by both the Auckland Chamber of Commerce and the Auckland Museum Council.[27] In 1945 it was suggested that the endangered mill be relocated to a reserve fronting on Beckham Place, on the other (eastern) side of Grafton Gully. But a number of residents petitioned the city council against this idea, complaining that it would interfere with their sea and harbour views.[28]

POETRY IN MOTION

Australian-born novelist and poet Eve Langley came to New Zealand in 1932, and for a time was living in one of the buildings alongside Partington's Mill. Auckland-born writer Ruth Park visited her friend Eve at the mill, located on the 'bony ridge' of Symonds Street. She described the building as having a 'farouche [sullen] reputation' and a 'romantically Bohemian' aspect. Further, she thought of it as 'a Tower of Trebizond' [in Turkey] or, recalling a German fairy tale, 'Rapunzel's prison'. She also likened it to 'a prodigious bottle tree, bulging at the base, tapering to two or three scraggy branches'. At that stage it only had two of its original four sails – a result of the recent fire; these hung in splinters and uttered 'a ghostly croak and flutter'. The base of the mill was surrounded by boxes, broken barrels, 'piles of rotting flour bags', and several buildings which appeared abandoned. It seemed deserted, and smelled of decayed wood and fermented grain. Entering the 'dank, cold and Gothic' mill, Park found herself in a room full of machinery, and millstones, 'enormous, prehistoric, powdered palely with husks of wheat'.[26]

The landmark is demolished

Despite the attempts of the Old Windmill Preservation Society, John Barr's fears of an 'irresponsible historical loss' were realised when demolition of the landmark began in April 1950. One positive outcome was increasing recognition of the need to preserve this country's heritage buildings, and in July 1952, when the National Historic Places Trust Bill was introduced in Parliament, reference was made to the 'public agitation' to preserve Partington's Mill. Unfortunately, 'so little interest was taken that that one-time landmark in the skyline has been lost'.[29] When the Bill was discussed again the following year, a Member described 'the old flourmill on top of the hill in Auckland' as the only one of its kind in New Zealand. He considered it unfortunate that some organisation had not been able to take it over and preserve it, whereupon the member for Grey Lynn reminded him: 'It was on offer.'[30]

By the 1970s, the site once dominated by the mill was occupied by Seabrook Fowlds Motors, the Maple Furnishing company and, on the corner of Symonds Street and Karangahape Road, the Caledonia Hotel. These, too, were all swept away and replaced by the Sheraton Auckland hotel, which opened in early 1983. Around this time the Auckland Civic Trust received a bequest of $25,000 from Parnell resident Mrs Myrtle Miller to go towards building a replica of Partington's Mill. It was to be erected within a kilometre of the original site, and the total estimated cost of the project, had it proceeded, was $570,000.[32]

In 2005, and under new owners, the Sheraton was rebranded the Langham Hotel Auckland, which retains a link with the past with the name of its restaurant: Partington's. In another connection with local history, Charles Partington's daughter Maria married David Goldie, timber merchant and later mayor of Auckland (1898–1901), and one of their five sons was the well-known painter of Maori portraits, Charles F. Goldie.[33]

RESTORATION PLANS

Although demolished, the windmill on Symonds Street remained on Auckland's conscience and was the subject of several restoration plans, all unrealised. In 1978 a Partington Mill Restoration Society planned to build a replica – subject to town-planning approval – and hoped to raise the estimated cost of $160,000. A spokesman saw such a building as a tourist attraction, and it was hoped that enough money could be raised to provide a bakery and tea rooms as well. Coincidentally, at the same time the Museum of Transport and Technology was considering building a copy of the old windmill. It was also pointed out at that if the original (brick) mill had somehow survived, it would now be on the list of earthquake-risk buildings.[31]

The Gates of Dawn - Auckland, N.Z.

7
ST PAUL'S

ECCLESIASTICAL AND HOMELIKE:
ST PAUL'S

LOCATION CAN BE AN IMPORTANT FEATURE of a building, but in several instances it has brought about their downfall. In 1885 Auckland's earliest church was demolished because the soil beneath it was needed for harbour reclamation. And as a visitor noted at the time, Aucklanders didn't appear too bothered by the disappearance of one of their most prominent landmarks. Further, this building had the added distinction of being designed by New Zealand's first architect.

The first church to be built in New Zealand was at Paihia, Bay of Islands, by missionary Henry Williams in 1823. It was constructed of raupo and served for five years until it was replaced; the present structure (built in 1925) is the fifth on the site. Auckland's first church dates from 1841, and was designed by William Mason. Born in Ipswich, Suffolk, in 1810, he trained as an architect and was responsible for several churches in Essex. In 1838 he and his wife and their young son sailed for Sydney, New South Wales, where he was offered the position of superintendent of public works in New Zealand. The family crossed to the Bay of Islands, and Mason was a member of the founding party that arrived at the site of Auckland in mid-September 1840.[1]

First Anglican parish

The Metropolitan Church of St Paul (hereafter St Paul's) in Emily Place was the first parish of the Anglican Diocese of Auckland. It was one of Mason's first projects in the town, and became its 'major architectural feature'. The foundation stone was laid on 28 July 1841, following a procession that had set off from Government House. It was led by Auckland's first (European) settler, and included a guard of honour from the 80th Regiment garrisoned at nearby Fort Britomart, Freemasons in regalia, the chaplain (the Revd J.F. Churton), and a large number of citizens and Maori. Governor William Hobson performed the ceremony, and beneath the foundation stone was placed a sealed bottle containing coins, early newspapers and the names of local individuals involved with the founding of the church.

ABOVE: *John Kinder,* Old St Paul's, *Auckland, 1861, watercolour, Auckland Art Gallery Toi o Tamaki.*
PREVIOUS: *Alfred Sharpe painted this early-morning view overlooking the Waitemata Harbour shortly before St Paul's was demolished.*

Limeburners, bricklayers and builders, excavators, fencers and brickmakers were invited to tender for its construction, which would consume an estimated 1000 bushels of lime and 300,000 bricks. The total cost of the church as built by Mason, but before the erection of the steeple, was a little short of £2918.

St Paul's was described as being 'in the old English style of architecture', and it bore a strong resemblance to one of Mason's earlier churches, St James', Brightlingsea (near Colchester, 112 km from London), built with Suffolk white brick and in Gothic style.[2] In both churches the main entrance was to the side of the building, in a tower with a square base and an octagonal steeple, while the end wall of the main building featured three tall and narrow arched lancet windows.

The first sermon

Bishop Selwyn, the first Bishop of New Zealand, preached the first sermon in St Paul's on 7 May 1843. The church was consecrated on St Patrick's Day, 17 March 1844, and served as the Anglican Diocese of Auckland's cathedral for over 40 years.[3] Selwyn's chaplain William Cotton was an early critic of aspects of the building, not appreciating that the roughness of the bricks inside would provide an ideal key for later plastering; nor was he able to visualise the final effect of the steeple, completed in 1844. Mason left Auckland for Dunedin in 1862, and the following year St Paul's was extended with a

> **'TRAGEDY ON SITE'**
>
> During the construction of St Paul's the architect William Mason and his wife suffered a great personal tragedy when the body of their nine-year-old son William was discovered in a well sunk on the site. Noted the *Otago Witness* on 1 July 1897: 'In connection with the construction of St Paul's a melancholy circumstance has to be recorded. There was no water on site, and to supply the deficiency Mr Mason had a well dug. One evening his only child … was missing from his home, and the next morning the boy, who was then about 10 years old, was found head first in the well.' No evidence of foul play was found, but there was a suspicion the boy had been murdered by one of Mason's workmen, who was later convicted and hanged for the attempted murder of a woman, a completely separate matter.[4]

new nave and chancel, completed to the 'very sympathetic' design of Colonel Thomas Mould, of the Royal Engineers, at a cost of £2500.[5] It became the garrison church while troops were stationed in the town, and it was claimed that every governor of the colony, when in Auckland, had worshipped at St Paul's. During the Northern New Zealand War the building was strengthened and loopholed (given vertical slits to allow the use of firearms) and designated a refuge for women and children should Auckland come under attack by Maori led by Ngapuhi chief Hone Heke.[6]

In 1860 Bishop Selwyn held his final service in New Zealand at St Paul's, and immediately afterwards boarded his ship and left for London.[7] In the late 1870s the interior of St Paul's was cemented, the churchyard fenced (at a cost of £700), and improvements made to the organ.[8] Around this time a local newspaper marvelled at Auckland's progress in general: tea-tree, scrub and raupo swamp had been replaced by paved roads, and nikau-covered whares superseded by palatial buildings 'constructed under the most approved architectural rules'. The foreshore was also changing, with reclamation in Mechanics and Freemans Bays and, more recently, the cutting down and 'tumbling' into the sea of Point Britomart.[9]

A huge landslip

Pressure on infilling the harbour between Queen Street Wharf and Point Britomart continued with the need to accommodate a railway line,[10] and was assisted by natural events. In 1883 a huge landslip caused some 2000 tonnes of debris to come down off the 50 m cliff of Point Britomart, causing much destruction to property and narrowly missing a passing cabman.[11] The slip led to discussions between three bodies with an interest in cutting down Point Britomart: the city council, which wanted to provide for an easy eastern exit from the city, the harbour board, which needed an alternative

St Paul's Church in 1883, two years before it was demolished. To the right, and on the corner of Princes Street and Eden Crescent, is the Auckland Museum.

A WELCOME SIGHT

St Paul's location overlooked Official Bay, and was a welcome sight for missionary (later Bishop) John Coleridge Patteson, who arrived in New Zealand in July 1855. He recorded: '[Auckland] looks much like a small sea-side town, but not so substantially built … rude warehouses, being mixed with private houses on the beach … on the west side is St Paul's Church, an Early English stone building, looking very ecclesiastical and homelike.'[12]

access to the harbour, and a meat-freezing company, which wished to use the earth for reclamation purposes.[13]

In its need for spoil, the meat-freezing company had arranged to purchase and demolish another nearby promontory, St Barnabas Point (near present-day Augustus Terrace, Parnell), believing — with some justification — that St Paul's Church would oppose the removal of Point Britomart.[14] But legal problems relating to the destruction of St Barnabas Point increased pressure on what remained of Point Britomart. The harbour board then took over the contract, ascertaining that there was 'nearly enough stuff' available at Point Britomart (an estimated 160,000 m of earth) to complete the job. It was a case of killing two birds with one stone; the removal of 'a very unsightly projection' and the completion of the planned reclamation.[15]

What remained of Point Britomart was seen as an 'obstructive mound' and a 'grievous eyesore'. Its complete removal down to the level of Fort Street would provide easy access to such parts of the city as Official Bay and Eden Crescent, which were otherwise cut off or only had 'roundabout access'.[16] All owners of private property affected by the proposed removal of Point Britomart proved 'reasonable' to deal with, recognising it as being good for business.[17] A number were happy to allow the work to proceed without making any claim for compensation, on the basis that their resulting lower and level sections would be worth far more than when situated on the edge of an embankment.[18]

A special case

But St Paul's Church was a special case. It stood directly in the line of the proposed development, and therefore needed to be removed. There was a suggestion that the attitude of the vestry of St Paul's had 'an aspect of obstructiveness'.[19] With the proposal to extend Princes Street in a straight line sloping down to the harbour, one possibility was to leave St Paul's standing in the middle of the street, with additional retaining walls built along its back and sides.[20] But an engineer believed this would endanger the building, so was unacceptable. The church then appointed a committee to consider the sale of the property, a move described as acting 'most judiciously, and in the most liberal spirit'. This led to discussions on an alternative location in compensation for the present building, and a majority of church authorities felt that an exchange of sites, along with a contribution of £3000 towards a new building, would be considered fair.[21]

The *Auckland Star* supported the removal project on the basis of economic expediency, noting that the buying-up of the necessary properties would never be cheaper. By way of example, St Paul's was 'old' and would therefore need to be replaced 'very soon' by a much more expensive building. Similarly, other buildings at the top of Shortland Street were 'among the oldest in the city' and

LOST HEADLAND

Point Britomart was a dominant headland on the edge of the Waitemata Harbour. In pre-European times it was occupied by Maori, and it was the site where the first Union Jack was raised in Auckland when the settlement was founded on 18 September 1840. The point became the site of Fort Britomart, one of the first British military fortifications in the colony, and, later, St Paul's Church. But beginning in the 1870s, Point Britomart was quarried away to provide fill for harbour reclamation, and nothing now remains of this once prominent landform. The continuing removal of Point Britomart led one citizen to complain that 'Every old landmark in Auckland is being ruthlessly swept away, and nothing left to mark the spot endeared by many old and varied associations.'[22]

would need to be replaced by 'substantial modern structures'. Therefore, such properties that stood in the path of progress needed to be bought up now, or else the later cost of compensation would be prohibitive.[23]

Demolition in a month

In February 1885 St Paul's was sold for £164, to be removed within one month, while its fences and gates went for £12 5s.[24] The final services were held in the church on 22 February 1885; a congregation of 22 attended an early communion service, with another 84 at midday, and there was a packed congregation in the evening. Demolition could now begin.[25]

English historian and novelist James Anthony Froude was visiting Auckland at the time, and observed the removal of Point Britomart and its church: 'Great works were in progress; labourers were swarming like bees, cutting away a huge projecting cliff to enlarge the area of the port. Bishop Selwyn's church – the first built in New Zealand – stood on the top of the precipice, and we arrived just in time to see the roofless walls before they disappeared in the falling rubbish. In a few days the church was gone. Sentiment belongs to leisure, and in the colonies, just now, they have none of either.'[26]

On 25 March workmen discovered St Paul's' foundation stone and beneath it a broken bottle containing coins and copies of early newspapers, but a parchment recording names was so shrivelled and faded that the writing was undecipherable.[27]

New church on a new site

The demise of Point Britomart also required the removal of the Hebrew Synagogue, which was provided with a new location in Princes Street. St Paul's was offered a new site on the corner of Symonds and Wynyard Streets, which was considered more central in an expanding city. In the meantime, a temporary wooden church, designed by William Skinner, was erected on the corner of Short Street and Eden Crescent.[28] The foundation stone from what was now known as Old St Paul's was relaid for its replacement, also designed by William Skinner, at Symonds Street on 11 June 1894. Although it lacks a planned corner tower and steeple, the present – and now third – St Paul's is considered a particularly fine example of Gothic Revival architecture.[29]

After relocating to Dunedin, William Mason was responsible for another church dedicated to St Paul, in Oamaru. It was designed to be built in two stages, but the congregation did not appreciate the first stage. As a result, and like St Paul's in Auckland, it was pulled down.[30] As the first architect to live and work in this country, Mason was responsible for the design of some of its earliest buildings. Many of them set standards in their day but are no longer with us, although notable survivors include his Old Government House in Auckland and St Matthew's Church, Dunedin.

8

VICTORIA ARCADE

'THE FINEST ERECTION IN AUCKLAND':
VICTORIA ARCADE

IN 1883 CONSTRUCTION BEGAN on what was confidently claimed would be one of the finest buildings in Auckland. It was dogged by controversy at the outset, and its demolition was lamented some 95 years later. This spectacular essay in red brick had a rich history, and enjoyed strong connections with the cultural life of the city. But it was allowed to deteriorate, and in 1978 the Victoria Arcade had the dubious distinction of being Auckland's first major building to be demolished in recent times. Its unnecessary departure was an unheeded warning of the destruction that would take place in central Auckland during the following decade.

In February 1883 the Auckland City Council put up for auction an inner-city section consisting of nine plots of land on 50-year leases. Bordered by Queen, Shortland and Fort Streets and located between the Post Office in Shortland Street and the Custom House in Fort Street, the site was advertised as being in 'the best business part of the city'. The council therefore demanded three-storey brick buildings 'of an approved style and character' which were to be erected within 12 months from 1 August next.[1] The property was currently occupied by five large brick buildings, including the Waverley Hotel and various 'unsightly' wooden buildings,[2] and the proposed redevelopment plan was seen as a step in the direction of Auckland having premises to compare with those in London's Threadneedle Street.[3] Tenders were called for the removal of the existing buildings,[4] and by October 1883 the site was cleared and ready.[5]

The New Zealand Insurance Company purchased all nine lots, and now planned one large building, the Victoria Arcade, to occupy the entire site. The building was the subject of debate even before construction began. There were suggestions the site should be retained as a public square, with one advocate for open space ('Anti-Goth') pleading: 'Have our city councillors no bowels of compassion?'[6] The official response was that the Arcade buildings would be 'a greater ornament to the city than the square', while the latter could be located directly in front of the Town Hall. The corner of Queen Street and Grey Street (later Greys Avenue) had been identified as a suitable spot for the Town

ABOVE: *View looking north down Queen Street with the tower of the Victoria Arcade building dominating the Shortland Street corner.* PREVIOUS: *The Victoria Arcade building is on the left (east) side of Queen Street, opposite the New Zealand Insurance Company building (demolished in 1915), and Partington's Mill is seen in the middle distance.*

Hall as long ago as 1872, but it would be 1911 before the building was officially opened. There was also a purely financial reason for not retaining the lower Queen Street site as a public square; the council's loss of rental income would be £3300 per annum, equal to a rate of 3d in the pound on the whole city.[7]

A handsome structure

A design competition was won by English architect Alfred Smith, and construction of the Arcade began – at a time when Auckland's single- or two-storey buildings were giving way to 'edifices of a more imposing character'. By November 1883 foundations were well under way for what was being hailed as 'the most extensive … valuable and

HERE FOR HIS HEALTH

English-born architect Alfred Smith was one of the designers of the Venetian Renaissance-style Army and Navy Club in London's Pall Mall, which opened in early 1851. He came to New Zealand for health reasons, and his best-known building, the Victoria Arcade, followed the style of one of his earlier projects, a grand mansion in Halifax, Yorkshire, whose Gothic style was also to be found in the design of the Arcade.

handsome structure in Auckland'.[8] But while the competition designs had called for the cost of the building itself not to exceed £25,000, the lowest (and accepted) of the five tenders was massively over budget at £36,162.[9]

By the following August the building's stone basement was complete, and the brick walls of the upper floors were taking shape,[10] but there was a succession of problems on site. A crane fell on a worker, rendering him 'insensible', although the building company was found not liable when sued by the aggrieved employee.[11]

In another incident, a carpenter was knocked unconscious by a falling board, suffering a fractured skull, but, contrary to all expectations, he recovered.[12] Contractors also had problems working within the small space enclosed by the hoardings around the site, but their complaints to the council went unheeded.[13] And following the threat of a reduction in carpenters' wages, 100 workers in Auckland – including 45 at the Victoria Arcade site – went out on strike. A 'vigilance committee' patrolled the Arcade, but were unable to prevent two men from sneaking in early one morning and starting work.[14]

In early 1886 the Auckland Institute of Architects arranged a collection of photographs of major buildings erected under their 'superintendence' to be displayed at that year's Indian and Colonial Exhibition in London. The Victoria Arcade was not among them, but it was in good company, for also excluded were the Supreme Court, St Andrew's Church, the Bank of New Zealand, and the Art Gallery and Library. As the *Auckland Star* pointed out, the Arcade's misfortune was that Alfred Smith, F.R.I.B.A., was not a member of that 'august body' the A.I.A.[15] But despite this apparent snub, no less than the Bishop of Auckland, William G. Cowie, believed that Smith's Victoria Arcade, along with his other buildings, bore 'the marks of a master hand', and were an 'ornament' to the city.[16]

Attracting retail tenants

By April 1886 the building was almost complete, and was hailed as 'undoubtedly the finest erection in Auckland'. In anticipation of it becoming a centre for the city's retail business, many of the shop and office spaces had already attracted tenants.[17] Shop premises had frontages to the street, as well as to the Arcade itself, while cellars and fireproof strongrooms were also available. Offices on the upper floors were advertised as suitable for solicitors, architects and surveyors, and accessible by staircases as well as a patent hydraulic lift – said to be the first in Auckland.[18] The first premises to be let were on the corner of Queen and Fort Streets, and taken by a draper for £9 per week for a five-year lease.[19] Other early tenants were the Board of Education and the American Consulate, although some two weeks later the United States Vice Consul died suddenly on the premises.[20]

By early June the building was receiving its finishing touches in the form of a verandah, which was designed to be supported from above. This was in case of the failure of its massive cast-iron columns, perhaps as a result of the impact of a runaway horse and cart.[21] Fire soon proved another threat, following a blaze in a shop caused by a defective chimney. Although it was quickly extinguished, there were concerns that had it occurred after-hours the new building would have been seriously damaged, 'if not destroyed'.[22]

A large domed tower

The Victoria Arcade had a total area of about 7280 sq m, of which the basement occupied 817 sq m.[23] To its street frontages the building presented three largely identical and symmetrical faces, combining the vertical lines of the narrow windows with the strong horizontals of the brickwork, all surmounted by gabled roofs with alternating pointed or rounded ends. Off-centre bays marked the entrances to the Arcade proper, while the whole richly ornamented edifice was dominated by a large domed tower near the corner of Queen and Shortland Streets. In 1887 the adjacent Shortland Street Post Office (demolished in the 1920s to create Jean Batten Place) considered asking the owners of the Arcade whether this 'conspicuous tower' could be used for the hoisting of a flag to indicate the arrival of the mail, as was the practice in Sydney and Melbourne.[24] In addition to the ground-floor shops, the building had 22 rooms on each of its three main upper floors, with another nine on the fourth floor in the gabled roof spaces, which were favoured as artists' and photographers' studios.

The Arcade quickly became a creative hub for Auckland. Artists who took studios in the new building included Robert Atkinson, Charles Blomfield and Louis J. Steele, followed by brothers Frank and Walter Wright, while Charles F. Goldie, Kennett Watkins and Edward Payton may have also had studios there at some point.[25] From 1886 a group of artists known as the Mahlstick Club (named after the painter's stick used to support the hand) met in Steele's studio, where they worked from models.[26] Others taking rooms in the building soon after it opened included the Auckland Political Financial Reform Association, the Auckland Benevolent Society, the Auckland Anglican Diocese, the board of governors of the Auckland College and Grammar School, and the board of Education (which also held examinations for senior scholarships at its offices).[27] The Auckland Athenaeum, dedicated to the pursuit of culture and science, rented and furnished a large room on the first floor of the Arcade, and made 'special provision for the accommodation of ladies'.[28]

ARTISTS IN RESIDENCE

The best-known artists known to have had studios in the Victoria Arcade were Charles Blomfield and Louis J. Steele.

Charles Blomfield was born in London in 1848 and came to Auckland in 1863. A self-taught artist, he was known particularly for his views of the Pink and White Terraces, which he first began painting prior to their destruction in the Tarawera eruption of 1886. His landscape paintings were in demand, and at one stage his studio was a tourist attraction. He built his own house in Wood Street, Ponsonby, and died in Auckland in 1926.

Louis J. Steele was born in Surrey, England, in 1843. He studied art in Paris and Florence, and came to New Zealand around 1886. By the following year he had taken a studio in the Victoria Arcade, and begun exhibiting at the Auckland Society of Arts. He painted portraits, historical subjects and racehorses, and also taught. One of his pupils was Charles F. Goldie, with whom he painted the popular *The Arrival of the Maoris in New Zealand* in 1898. Steele died in Auckland in 1918.

Because of its central location, shop windows in the Arcade were used for displaying items of public interest. In November 1886 photographers Burton Bros displayed a piece of 'geyser pipe' ejected by the recent eruption of Mt Tarawera which had been collected by Charles Blomfield. Edward Payton, who shortly became director of the Elam School of Art, displayed etchings of New Zealand scenery at Wildman's bookshop,[29] and when the United States declared war on Spain in 1898, the 'Stars and Stripes' was patriotically displayed in the window of the American Consulate.[30]

During the period 1914–18 the Victoria Arcade provided temporary accommodation for the staff of its owners, New Zealand Insurance, whose new building was being constructed on the opposite side of Queen Street. By 1925 the Arcade was occupied by no less than 14 architectural practices, among them Gerald Jones, who specialised in large Arts and Crafts-influenced houses, and H. Clinton Savage, responsible for the George Court's Building (1924) in Karangahape Road.[31] By the early 1930s the building was still home to Wrights Studio (although Walter Wright would die in 1933), as well as artist Minnie White and, as usual, several architects.

Artists and exhibitions

For a quarter of a century the Victoria Arcade was also home to the Auckland Society of Arts. From 1905 the society had its own building (designed by architects Goldsbro' and Wade, also based in the Victoria Arcade) in Coburg Street (later renamed Kitchener Street), which it sold in 1926. It continued to rent its old building for large and annual exhibitions (which were later moved to the Auckland Art Gallery), and around 1930 it took clubrooms in the Arcade. These provided a comfortable meeting place in the city for members, with a well-stocked library of art publications, and were used until the society moved to its own premises in Eden Crescent in 1956. The clubrooms in the Arcade

The rich texture of the brick façade of the Victoria Arcade is captured in this c.1982 photograph by John Fields.

ARCHITECTS AND PHOTOGRAPHERS

In 1902 the *Cyclopedia of New Zealand* described Auckland's architecture as 'decidedly imposing', and listed (in this order) the following as examples of buildings that were 'all handsome, and some of them noble': the Victoria Arcade, Supreme Court, Customs House, and City Library and Art Gallery.[32] It is likely that a significant number of Auckland's other buildings were designed in the Victoria Arcade, given its strong attraction for architects. In 1912 there were eight architectural practices on the premises, along with several surveyors and well-known photographers Henry Winkelmann and Josiah Martin.[33] The architects included brothers Norman and Henry Wade, the first of whom was later involved (with Alva Bartley) in the design of several significant local landmarks, among them the modernist Broadcasting House (1933) in Durham Street, which was inexplicably demolished as recently as 1990.

were also a regular venue for smaller and one-person exhibitions, and among those who showed work here were painters Theo Schoon, Eric Lee-Johnson, Rex Fairburn, E. Mervyn Taylor, Dennis K. Turner, Ron Stenberg, John Weeks and Louise Henderson, and sculptors Molly Macalister, Alison Duff, Greer Twiss and Jim Allen.[34]

After years of accommodating artists and architects, along with its fair share of solicitors and accountants, the early 1940s saw organisations of a rather different nature gravitate to the Arcade. Rooms were now taken by the Rosicrucian Order and the Anthroposophical Society and, as if to provide balance, the Rationalist Association.[35] The latter was another long-term tenant at the Arcade, based there from the early 1930s until 1960, when it purchased its own property, the sensibly named Rationalist House, in Symonds Street.

Demolish or redevelop?

But by the early 1970s the Victoria Arcade had lost its original sparkle. It became even more Gothic in one sense, with its brickwork having turned to a dull red, and its stonework darkening to a dull grey. By now it had long ceased to operate as an arcade, while its once distinctive corner tower had been removed and its original verandahs replaced.[36] In 1974 the Bank of New Zealand bought the lease, but at that point was undecided as to whether it would demolish and redevelop the property. It was a requirement of the lease that if a new building was not completed on the site by 1982, the present structure would revert to the owner of the land, the Auckland City Council.[37] In 1975 the Victoria Arcade went on the market for $1,450,000, and its days were numbered. The following year architectural historian John Stacpoole noted that, despite its interesting features, the building's 'drab colour and impractical planning' had not endeared it to the public.[38]

Demolition bricks in demand

In October 1977 notices to quit the building by the following April were served to tenants, including the resident caretaker, who occupied a large four-bedroom flat on the fourth floor. Among those now needing to find a new home were the New Zealand Foundation for Peace Studies, Birthright, Friends of the Earth, Pregnancy Help, the Gay Publishing Collective and – the longest-serving tenant – Worrall Jewellers Ltd, manufacturing jewellers and engravers, who had been there for 77 years.[39] By early October 1978 the wrecker's ball had reduced the Victoria Arcade down to the first-floor level. Nearly half of its estimated one million bricks were for sale, at 15 or 20 cents each depending on their condition, and were eagerly sought by home improvers engaged in the current vogue for laying patios or building backyard barbecues.[40] It was also reported that the demolition of the building had been eagerly anticipated by local boat-builders keen to get their hands on its 30 m-long kauri beams.[41]

A late 1880s view of the Victoria Arcade building on the corner of Fort (left) and Queen (right) Streets.

Demolition was complete by mid-October 1978. To add insult to injury, the Victoria Arcade would now be replaced by a decidedly undistinguished Bank of New Zealand building. However, it would have been of some satisfaction to Alfred Smith's Gothic arcade that its replacement lasted a mere quarter of a century before it, too, was cast aside and superseded by the Deloitte Centre, completed in 2010.

9
HIONA

MYSTIC SIGNS AT MAUNGAPOHATU:
HIONA

THE HOUSE OF HIONA, at Maungapohatu in the Urewera country, was one of the most remarkable buildings ever erected in New Zealand. It was as remote as it was short-lived, and owed its origins to a spiritual experience. Its end was also quite out of the ordinary, for while buildings have frequently fallen victim to redevelopment or natural forces, this one disappeared for purely philosophical reasons.

Following a revelation on Maungapohatu, the mountain sacred to the Tuhoe people of the Urewera, Rua Kenana Hepetipa (b. 1868/9) claimed to be the successor to the late Te Kooti Arikirangi Te Turuki, the Maori leader and the founder of the Ringatu Church. In 1906 Rua announced he would ascend the throne, and that Turanga (present-day Gisborne) would receive a visit from King Edward VII, who would assist Maori in their struggle to reclaim their traditional lands from the encroaching Europeans. When His Majesty failed to appear as predicted, Rua advised that *he* was in fact the king, and set about establishing a new community at the foot of Maungapohatu. He drew heavily on scriptural history; his followers were called Iharaira (Israelites) and, like the Nazarites in the Hebrew Bible (Numbers 6: 1–12), they abstained from alcohol and allowed their hair to grow long. Rua also followed the example of the polygamist King Solomon of Israel and by 1908 had seven wives.

The site selected for the community lay on a proposed stock route linking Ruatahuna and Gisborne. The bush was cleared, and buildings were constructed from pit-sawn timber. Pastures were established, and the north-facing and well-drained slopes proved ideal for sheep and cattle. Progress was widely reported in the New Zealand press, and the admiration for Rua's achievements was apparent in an account published in the *Poverty Bay Herald*: 'In selecting a site for his township, Rua was not afraid to face difficulties. Instead of choosing some flat and unhealthy low lying ground, as the Maoris generally do, he attacked the side of a rising piece of land, and formed a great artificial sidling [an area of sloping land, or hillside].' Already over 500 hectares of bush had been felled, and the new settlement's utilities included water from several springs, which had

ABOVE: *View of the settlement, Maungapohatu, 1908, with Rua Kenana's temple, Hiona, on the right.*

PREVIOUS: *Hiona, with the freestanding platform from which Rua Kenana addressed his followers, to the right.*

---

been channelled into a succession of pools reserved for specific functions, from cooking to washing and bathing. The streets were lit at night by kerosene lamps and there was an 'excellent' butcher's shop and a general store, both run by a committee.[1]

'A spot of unsurpassed beauty'

Dr James Bell, Director of the New Zealand Geological Survey, carried out field work in the Urewera. His exploits were published in 1914 and included a description of his encounter with Rua's settlement: 'Though we got a daylight start, it was ten o'clock when from a high point on the track we came suddenly in sight of the stronghold of which we had heard so much for the two or three days before. Rua had chosen a spot of unsurpassed beauty for his headquarters. It lay in a high and extensive terrace, situated at the base of densely forested slopes rising to the well-defined escarpment marking the crest of Maunga Pohatu … A few *whares* lay scattered near the river and on the lower terraces, but our attention soon wandered from those humble older dwellings to the imposing habitations of the prophet, which rose, fresh and clean, above the clearing.'[2]

Bell was obviously impressed by the industry of Rua's followers, whose systematic bush clearance was an improvement on the usual approach taken by settlers: 'It seemed scarcely credible that all the work which we saw before us had been done in a few months' time. To any one who knows the difficulties of clearing the New Zealand forest, the tremendous task of removing the great stumps of rata or the rimu, even after a forest has been removed, will be apparent … Rua had been unlike the average agriculturist – the cleared ground was almost entirely free from fern stumps, and there was no unsightly untidy area between clearing and forest. The newly turned ground began where the great virgin forest ceased, giving a particularly pleasing effect.'[3]

Maungapohatu's most impressive buildings

Surrounding Maungapohatu was a low stockade of split timber, and at the main entrance to the community was the sign 'Mihaia' (Messiah), announcing Rua's status. The settlement itself was clearly divided into an inner and outer area, with the former enclosed by a fence and including various important buildings and workshops, and upwards of 80 sleeping houses described by Bell as 'neat' and 'well-built [of] palings and shingles'.[4] Beyond, in the outer area, were the cook houses and store houses. But the most impressive buildings were those associated with Rua himself, situated on an elevated ridge in the inner area overlooking the settlement. As described by Bell, the temple was named Hiona (Zion), and was also referred to as the 'great court or meeting house'. The rotunda-like building was lined throughout and, according to the *Poverty Bay Herald*, the lower part of the two-storey structure measured 60 ft (18.3 m) in diameter – considerably less than had been recorded by Dr Bell – and was about 20 ft (6.1 m) high.[5] Some seven decades later, Peter Webster would reduce the diameter even further, to 12.19 m, while putting the overall height of the building at 10.67 m.[6]

Hiona's design

A likely precedent for the design of Hiona was the Dome of the Rock mosque, also known as the Temple of Solomon, in Jerusalem, while Rua had earlier built a small round house of split palings and canvas at his camp at Rangitata on the upper Hangaroa River, some 20 km east of Maungapohatu.[7] In terms of function, Hiona was also influenced by the New Zealand Parliament Buildings in Wellington. In 1854 this country instituted a bicameral legislature, consisting of a House of Representatives and a Legislative Council (or Upper House, which was abolished in 1950), a system followed by Rua. In fact, he intended his building to replace Parliament in Wellington which, coincidentally, burned down in 1907.

RUA'S TEMPLE

Dr James Bell described the two main buildings at Maungapohatu: 'At the south end of the roadway stood Rua's residence [named, Hiruharama *New Jerusalem*] – a two-storey, double-gabled building, with a corrugated iron roof. Near by were a number of other buildings … but by far the largest and most interesting building was the temple, standing near the northwest corner. It was a round building about 75ft (22.9 m) in diameter and 30ft (9.1 m) in height, surmounted by a cupola …'[5] Around the latter ran a narrow balcony, and from this an elevated gangway led to a freestanding tower, from which a stairway descended to the ground. During fine weather Rua delivered messages to the faithful from the tower; when conditions were unfavourable, he spoke from a raised platform in the centre of the temple.

Rua Kenana in front of a wooden turnstile at the entrance to Maungapohatu, 1908. Above him is the sign 'Mihaia' (Messiah) and images of a star, playing-card symbols and comets.

Hiona operated as a council chamber, with an upper and lower house. At the centre of the lower hall was a large table where a committee of 12 chosen by Rua, and referred to as the 'apostles', sat during court sessions. They listened to the evidence of the accused (whom they did not see) and, without speaking, recorded their opinions on paper. The table was rotated to allow each juror to add his views in turn, and when all had done so the papers were carried up to the upper house. This storey measured about 6.1 m by 3.6 m and had a large central circular opening guarded by an ornamental railing. It was reported that Rua intended some sort of device to hoist the papers up to him through this opening, but no such system was in place by the end of 1907. After examining all the evidence delivered from below, Rua and two assistants then pronounced judgment.[9]

Dr Bell described the temple as 'quite a remarkable building', and the upper house as 'a sort of holy of holies' to which only Rua and his apostles had access. The building was also 'gorgeously painted in yellow, blue and white, with Mosaic [referring to Moses] designs which are supposed to have some symbolic meaning'.[10]

The exterior of Hiona was clad with white-painted vertical boards decorated with yellow diamond and blue club motifs taken from playing cards. These images were commonly used in the 19th century as mnemonics (memory aids) to explain the Scriptures to those unable to read, and Te Kooti had earlier devised such a system for the Ringatu Church. For the residents of Maungapohatu, the playing-card club symbol on Hiona represented the King of Clubs, the king who is yet to come, with whom self-proclaimed prophet Rua identified. The diamond motif stood for the Holy Ghost and related to Rua's revelation on the mountain, while the colours blue and white were associated with ancient Israel.[11] There were also 'mystic signs' at the entrance to Hiona, described as being 'in poor imitation of the square and compass of Freemasonry, with Rua's favourite insignia, a double-tailed comet'.[12]

Christmas 1908 at Maungapohatu was marked by a gathering that included some 200 of Rua's followers, as well as Pakeha visitors. Among other things, the *Poverty Bay Herald* was impressed by the dawn chorus in the Urewera: 'Christmas morning was ushered in with prayer in the open before the dwelling of the high priest. The silvery notes of an invisible feathered choir in the neighbouring forest blended quaintly with the harmonious chanting of the congregation, while the solemnity of the service, combined with the strange appearance of the long-haired worshippers, contributed greatly to the impressiveness of the scene.' The same reporter noted that the settlement itself had not changed much over the previous year, although the surrounding cleared land was now 'verdant' with rye-grass and thriving crops of potatoes, maize and melons.[13]

Police invasion

By 1910 Rua had relaxed the rule requiring abstinence from alcohol, and he now came under suspicion for illegally supplying alcohol at Maungapohatu. Inquiries were made, and on 10 January 1911 six constables arrived at Rua's pa at Waimana, some 40 km north of Maungapohatu. He was arrested and received a suspended charge.[14] Government authorities had long regarded Rua as a disruptive influence, an attitude that hardened when he objected to Tuhoe volunteering for military service in the First World War. Rua's actions were considered seditious, and by May 1915 police were able to produce five more charges for sly-grogging. He was summoned to attend court in Rotorua, but refused, and in March 1916 an armed force was sent to detain him.

At Rua's Supreme Court trial the jury threw out the charge of sedition, but Rua was found guilty of resisting arrest and consigned to Auckland's Mt Eden Prison with the

THE ARREST OF THE PROPHET

On 2 April 1916 an advance guard of constables arrived at Maungapohatu and, following a scuffle during which shots were fired and two Maori (including Rua's son Toko) were killed, Rua was seized. The mêlée has been described as the worst clash between police and a Maori community in the 20th century. As reported in the press, Rua was 'overpowered and manacled... by the armed representatives of the law which he had defied'.[15]

excessively harsh sentence of one year's hard labour followed by 18 months' imprisonment. There were also controversial circumstances surrounding Rua's arrest, and individuals involved in the operation have since been accused of perjury and orchestrating evidence.[16]

Sweeping changes were now taking place within Maungapohatu. By 1914 these included the building of a new meeting house, and would lead to the reconstruction of the whole community. Hiona was falling into neglect; the exterior access to its upper storey had been removed, while the building itself became a general storeroom for the community's wool clip and stock of grass seed.[17]

Return to Maungapohatu

Following his release from prison in April 1918, Rua returned to Maungapohatu and was met by a large crowd. But having been humiliated by the New Zealand Government, he now determined to make a new start, and reject the failed ways of the past. The spilling of blood and the violation of Rua's tapu were factors behind his decision to begin what was effectively the third reconstruction of the settlement. He ordered the removal of the symbols of the old order, and the establishment of a new centre at Maai, a short distance below Maungapohatu in the Waikare River valley. Rua's own house was dismantled and the timbers used for constructing a new and smaller house (also named Hiruharama) for the prophet and his wives. Meanwhile, the timber from Hiona was used to build a new meeting house which opened in February 1919, and was named Te Kawa a Maui after a house of learning that once stood at Maungapohatu. The boards taken off Hiona were recycled at random, and while whitewash now obliterated the original coloured decorations, these later reappeared as the concealing top coat wore off.[18]

Rua Kenana died in February 1937 at Matahi, a community he had founded in the eastern Bay of Plenty in 1910. His meeting house Te Kawa a Maui survived another 40 years until 1977 when it burned down, thereby extinguishing the last trace of a remarkable building which had been inspired by a revelation on a sacred mountain.[19]

10

THE ROUND HOUSE

A LANTERN IN THE TOWN:
THE ROUND HOUSE

ONE OF THE MORE UNUSUAL of New Zealand's rich stock of demolished buildings was an early example of thinking outside the square. New Plymouth's so-called Round House was also distinguished by its prefabricated construction and intrepid journey from England. It began life as a family home, and was later commandeered for military purposes and as a boarding house. It was also extended and relocated, and remained a well-known local landmark until it was demolished, after some 86 years of service, to make way for a car sales yard.

In December 1851 Thomas Hirst, his wife, Grace, and five of their eight children sailed from England on the barque *Gwalior* for the New Zealand Company settlement of New Plymouth. In addition to the usual hazards at sea, they had to endure an alcoholic captain and a mutinous crew. In fact, the ship may never have reached New Zealand had it not been for one Chantrey Harris, a passenger who boarded at Cape Town and took charge when the captain and second mate drank themselves into a state of 'delirium tremors'. The *Gwalior* eventually reached Auckland and later sailed on to New Plymouth, where it arrived on 18 August 1852 and, despite a strong gale blowing at the time, managed to unload its passengers and cargo.[1]

Distinctive local landmark
Thomas Hirst had started business as a wool merchant in Bradford, England, and now established himself as a wool classer and buyer in New Plymouth. He also built the family's first home, the frame (which, according to one account, was of steel) and timbers for which had been precut and brought out from England, and 'floated ashore in bundles' from the *Gwalior*.[2] The house was erected on a section on the Devon Line (later Devon Street East), as the settlement's main street was then known. Originally of single storey, octagonal in plan and built around a central chimney, it soon became known popularly as the Round House, and later – perhaps more officially – as Egmont House. Its distinctiveness made it a local landmark; in 1868 a butcher advertised his shop as 'opposite the Round House'. Strangers to New Plymouth town could be directed by

ABOVE: *Bernard Aris,* The Old Round House, New Plymouth, *monochromatic watercolour, with ink, on paper. Bernard Aris (1887–1977) was born in Surrey, England, and came to New Zealand in 1922. He lived in New Plymouth and became a prolific painter, his output including over 600 sketches and watercolours of Mt Egmont/ Taranaki.* PREVIOUS: *The original location of the Round House: from Gover Street looking towards Mt Eliot.*

the instruction 'so many doors past the Round House', and at night when lit it had the appearance of 'a huge old stable lantern'.[3]

The Hirsts lived in the Round House until 1854 when they moved to a small farm at Bell Block, some 7 km north-east of New Plymouth. Their new property was called Brackenhirst, presumably in recognition of Grace Hirst's family name, Bracken, rather than the local vegetation. With the outbreak of hostilities, and the declaration of martial law in New Plymouth in 1860, the family moved back to the safety of the main settlement, and their Bell Block homestead was burned and largely destroyed. During the period of the Taranaki Land Wars the Round House was taken over by the military and used for 'conferences' and as a messroom, and also as a shelter for settlers made homeless during the conflict.[4]

In October 1862, after a visit to England, Thomas and Grace Hirst returned to New Plymouth and bought another property, Willow Field, where they built a row of houses for each of their children. The Hirsts reoccupied the Round House in 1869, and at some point extended it beyond its octagonal footprint and added an upper storey. In addition they built another house alongside it, also of two storeys but of conventional square design. The two structures were linked, and the extended building became known collectively as The Round and Square House.[5]

Unusual exterior cladding

The timber used in the construction of the original Round House was heart of red pine.[6] Like the later adjacent Square House, it was distinguished by its exterior cladding of vertical board and batten. This system was commonly used for cottages in New Plymouth at that time, and consisted of wide boards fixed vertically with the joints between covered by narrow battens. From the time of organised settlement from Europe, New Zealand house-building was dominated by the use of an exterior cladding of horizontal weatherboards, but board and batten was a simpler and common alternative, although in later years it was used mainly on outbuildings or cheaper houses.[7]

EARLY PREFABRICATED HOUSING

Imported prefabricated structures – like the original section of the Round House – were hardly unusual in colonial New Zealand. In the 1840s Britain led the world in the manufacture of such systems, which ranged from modest wooden huts to cast-iron villas and even hotels and churches. Two of New Zealand's earliest official buildings had overseas origins; Auckland's first Government House was made in England, while the heart of the Waitangi Treaty House was prefabricated in Sydney, Australia. Some houses were exported as complete kitsets, while others consisted of only a framework, which was later clad with locally available materials. While most of its imported houses originated in England, New Zealand was also enriched by examples from Tasmania, as well as India and France.[8]

The Hirsts lived in the Round House until Thomas's death in October 1883. He had taken an active part in local matters in New Plymouth, and was a member of both the House of Representatives and the Provincial Council, as well as a Justice of the Peace. He was also said to possess 'a genius for inventions'. He made models of his ideas, one of which was reportedly inspected and approved of by Sir George Grey during a visit to the town. Perhaps this particular model explained Hirst's novel scheme for landing passengers and cargo from ships at New Plymouth, and may have been inspired by the gale-force conditions when he first arrived there in 1852.[9]

Hirst's idea for New Plymouth, detailed in a model which had occupied him for two months, was to have two fixed iron stages, one at some 366 m above the high-water mark and another at the low-water mark. Carriages 1.8 m by 1.2 m by 1.8 m would be drawn between the two stages by a wire rope, powered by an engine on shore. Hirst proposed that the stages would be made out of old railway iron, which he understood could be bought cheaply in England.[10]

New owners

Following Thomas's death, Grace Hirst moved out of the Round House and the old family home now underwent a number of changes of ownership and use. At various times it operated as a boarding school, 'dancing seminary' and private residence, and was where Mrs J.W. Batten, organist at the Wesleyan Church, gave piano lessons. But it probably spent most of its remaining years as a boarding house.[11] In 1890 it was the scene of a fire, fortunately extinguished by the local brigade, and around this time the residents of the house regularly advertised for domestic servants.[12]

The year 1902 proved to be particularly dramatic for the property; in April it was sold for 'a satisfactory price' to a local investor, while an earthquake in December caused damage to its chimneys – as well as other buildings throughout New Plymouth.[13] The following year the new owner subdivided the property into four shop sites, to be sold at auction. Three of the sections had a depth of 40.8 m and frontage of 6.4 m to Devon Street, and one of these was occupied by the Round House, which the vendor was required to remove. The fourth section was slightly narrower and occupied by what was previously known as the Square House and now operated as the Strathmore Boarding House, suggesting that the two parts of the Hirst homestead were now separated. All of the sections were sold in five minutes at the auction, allegedly demonstrating 'how eager a demand there is for Devon Street and central property'.[14]

New location
As per the sale agreement, both the Round House and the Square House were moved back from Devon Street to a frontage on Courtenay Street, opposite the Central Infants' School. The relocating of this 'ancient building' was described as 'a slow and arduous pilgrimage', and it was hoped that in its new location it would be able to carry on its old duty as 'watch tower and guiding post'.[15] Shortly after the shift, a visitor exploring the house's 'odd nooks and corners and quaint triangular little rooms' found the walls to be in extremely sound condition, and suggested they would probably outlive many of those belonging to the 'modern up-to-date villas'. So tough was the timber that a contractor was unable to drive in a nail; when pounded with a hammer it failed to make an impression and simply buckled.[16]

The old Hirst home remained a local landmark; in 1905 a resident advertising for lost property gave an address as simply 'Square House, New Plymouth'. But in 1938 time ran out for the Round House. After beginning life on the other side of the world, and adjusting to a wide range of uses, this long-serving landmark was demolished. At the time it had two rooms on the ground floor and six bedrooms – four three-cornered and two rectangular – on the floor above. The structure's timbers were still in a remarkable state of preservation, and it had the original locks in the doors.

The Round House was demolished to make way for extensions to Len Nicholls Motors Ltd, which later gave way to Phillips Ford Motors and/or Moller Motors. But both of those are also now gone, having been replaced by the site's present occupant, the ubiquitous Warehouse.[17]

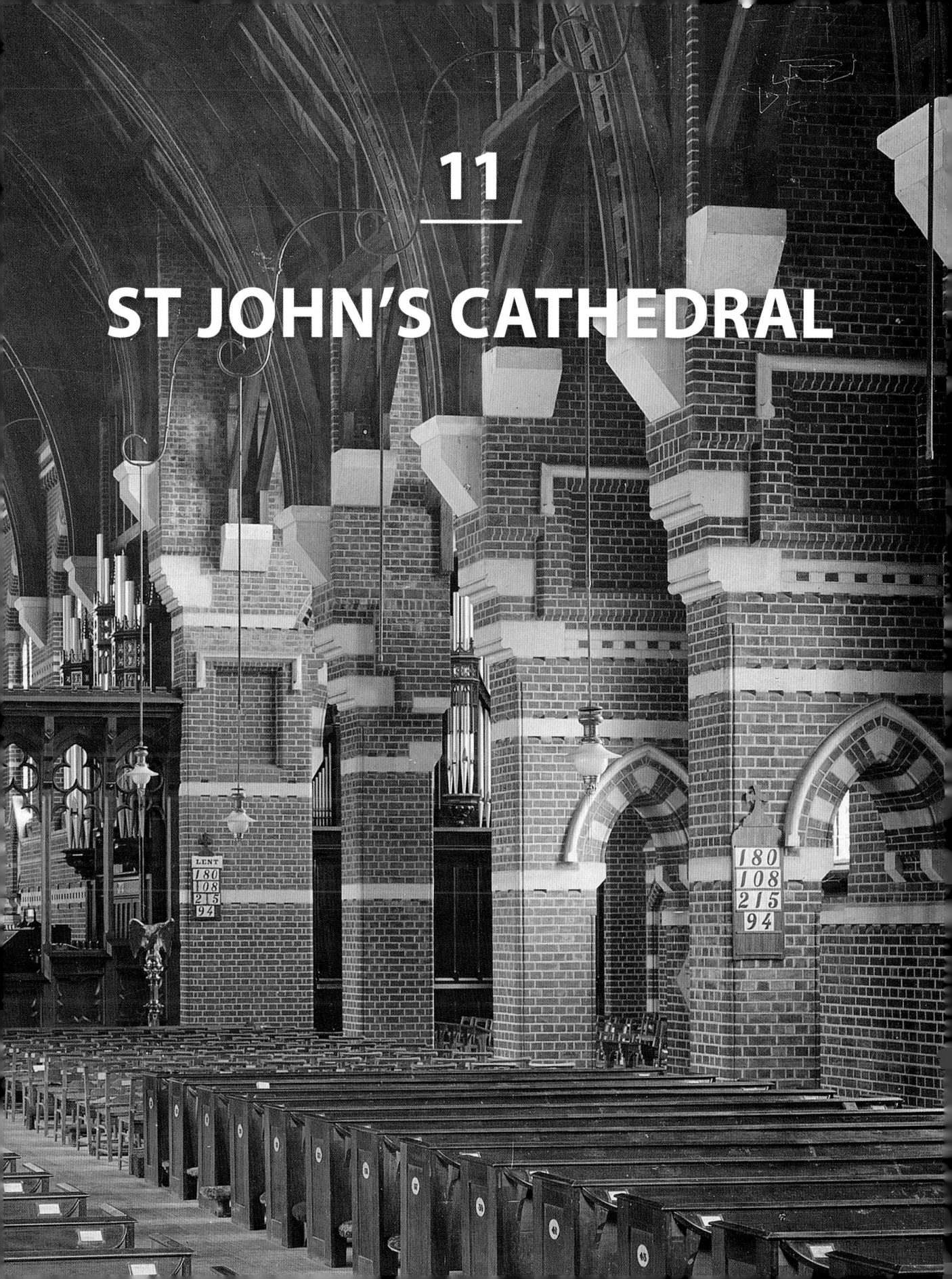

11
ST JOHN'S CATHEDRAL

PRIMITIVE GRANDEUR:
ST JOHN'S CATHEDRAL

In 1886 the Anglican community in Napier received its new cathedral, St John's, the largest brick building constructed in 19th-century New Zealand. It was the most ambitious design yet undertaken by leading architect Benjamin W. Mountfort, with an interior unlike anything else attempted in the country at the time. But on the morning of 3 February 1931 this grand building was reduced to what was described as 'a pitiful pile of bricks'.

The Church of England had its beginnings in New Zealand with the arrival of the Revd Samuel Marsden at the Bay of Islands in 1814. Bishop Selwyn, the first bishop of New Zealand, arrived in 1842, and in 1859 he ordained William Williams, bishop of the new Diocese of Waiapu, named after the East Coast river where the Anglican mission began in the 1830s. Following skirmishes between Government troops and members of the Hauhau movement and damage to his mission station, Williams moved his headquarters from Poverty Bay to Napier in 1867. Hawke's Bay, previously part of the Diocese of Wellington, was added to that of Waiapu in June 1869, and Napier became the cathedral city.[1]

Prior to the appointment of a clergyman, early Church of England services in Napier were conducted by a citizen. Bishop Selwyn visited the district at this time, and when asked to provide a clergyman he reportedly said he could do nothing of the kind, 'unless locals put their hands in their pockets'. They did, the necessary funds for a stipend were raised, and the first local Anglican clergyman was appointed. At first worshippers used a small wooden building, which later became a schoolroom, and this was followed by the first section of the wooden church of St John's. Over the next decade it was extended, and had a seating capacity of 600, but by 1885 it was in need of either extensive alterations or replacement. The decision was taken to build a cathedral, one in 'a beautiful ecclesiastical style', so Benjamin Mountfort was approached.[2]

ABOVE: *The ivy – (or Virginia creeper) – clad exterior of St John's Cathedral, Napier, c.1910.*

PREVIOUS: *Interior of St John's Cathedral, Napier, c.1890s.*

Most ambitious undertaking

In 1884 Mountfort used brick for the first time as the main material in a church, and this was a precursor to the largest such building he would design, St John's Cathedral, Napier.[3] By this time he had already carried out several church commissions in the Diocese of Waiapu, but the Cathedral was his 'most ambitious ecclesiastical undertaking'. It was all the more remarkable for being in a provincial town. (According to the 1886 census, the population of the Napier borough was only 7680, with another 1504 in neighbouring Hastings.) The Cathedral gave Mountfort an opportunity to develop earlier ideas on a much larger scale, and he visited England and Europe, presumably looking for innovations that might be suitable for New Zealand.[4]

The foundation stone for St John's came from Hoon Hay, near Christchurch, and was laid on 29 September 1886 by the Bishop of Waiapu. The consecration took place later than planned, on 20 December 1888, having been delayed on account of the bishop's visit to England.[5] The total (external) length of the building was 52.9 m, the breadth at the transept 26.1 m, and the height from floor to ridge 17.7 m. The foundations consumed

356.3 cubic m of concrete, there were some 99.1 cubic m of stone, 452,000 bricks, and some 33,200 m of timber (all from local forests) in the building. The total cost was £11,000.[6]

> **FOREMOST CHURCH ARCHITECT**
>
> Architect Benjamin Woolfield Mountfort was born in 1825 and grew up in Birmingham, England. In 1844 he began working under a Gothic Revival church architect who would have a major influence on his own design principles. In 1850, and by then a practising architect in London, Mountfort and his wife emigrated to New Zealand. They were among the first group of Canterbury Association settlers, arriving in Lyttelton on 16 December. He was soon working as an architect, and, in partnership with his brother-in-law, was responsible for the Canterbury Provincial Council Buildings (1858–65). From 1866 he worked on his own, as supervising architect for Christchurch Cathedral (from 1873), while other jobs included the Canterbury Museum (1869–82) and Canterbury College (1877–82). By the 1880s Mountfort was recognised as New Zealand's foremost church architect.

Honest use of materials

A reviewer of the recently completed St John's made the general observation that each architectural material had its own appropriate treatment, and should be used honestly. Wood, for example, should not be made to look – or act – like stone. The lack of local quarries prevented the Cathedral from being constructed of stone, but the extensive local clay beds suggested brick would become the preferred material for the province's permanent buildings. Some considered it an inferior material for such purposes, and there was the matter of its structural suitability given that Napier had experienced earthquakes in the past. However, certain brick buildings were recognised as ranking among the 'grandest and stateliest' in the world, while in parts of Germany and Italy the use of the material had been brought to 'a wonderful pitch of perfection'. And whereas a stone building might be distinguished by arches and curves, these necessarily adopted rectangular forms when made out of brick.

In the new Cathedral, the brickwork of the walls was relieved by stone string courses and flat bands of stone, while the bricks had been moulded in a range of patterns to produce an effective variety of detail. The new church was seen as 'more or less successful' in pointing the way to a faithful application of local materials. And although at that stage it was short of a tower, it was roofed in slate of two colours, arranged in patterns, forming a conspicuous feature and topped with a perforated ridge cresting.[7]

A soaring lofty roof

The same reviewer described the effect of entering the building: 'In one wide span of thirty-seven feet [11.3] m the lofty roof soars away for nearly one hundred and seventy feet [51.8] m in a long, solemn, stately perspective to the great eastern window; nothing breaks the continuity of its leading lines. The interior is unlike anything yet attempted in this country, and by its spacious effect reminding the spectator of a wide basilica.' There being no beams in the roof, the enormous pressure put on the walls by the arches was shouldered by large abutments, which were mostly contained internally. Double arches supported purlins, rafters and diagonal boarding, all of which was given lightness and interest by means of much pierced and perforated work.[8]

On either side of the north transept were windows from the old St John's. In general, windows were of many 'counterchanged colours and hues', producing a great range of effects depending on the position of the sun. While some had predicted that the interior of the church would be dark, the light was in fact subdued ('a most essential matter in our brilliant and trying atmosphere'), plentiful and 'of a proper, sober and solemn character'. To summarise, the finished effect demonstrated how the judicious handling of local materials could produce work 'not unworthy of the exalted purpose for which it was designed'. Even so, the same writer suggested local brick manufacturers could improve their wares, and not 'rest in mediocrity'. Reference was made to late 19th-century art writer John Ruskin, who believed clay was never intended to be made only into oblong shapes of one size. Instead, it was a material from which workmen might 'knead out some expression of human thought'.[9]

One hundred and eleven years later, art historian Ian Lochhead described St John's as 'a building of primitive grandeur' and 'poised at the beginning of a new tradition of church building'. For him the Cathedral's use of elemental forms and overlapping planes of brick represented a link with the monumental simplicity of medieval tithe barns. Mountfort had managed a skilful arrangement of bricks to produce a wide range of surface patterns, and achieved impressive effects 'with the simplest means', while the extremely spacious interior – which could accommodate 1100 people – was unlike anything yet attempted in New Zealand.[10]

Earthquakes strike

While the foundation stone for the Williams Memorial Chapel, on the east side of the Cathedral's north transept, was laid in 1886, it was not completed until 1902.[11] Two years later, on 9 August 1904, an earthquake was felt throughout Hawke's Bay, Whanganui and Wellington. Many Napier chimneys were toppled, houses reportedly 'rocked like ships in a storm', and there was also a major slip at Bluff Hill.[12] Elsewhere in the region, telegraph lines were brought down between Te Aute and Dannevirke, and

All that remained of St John's Cathedral, Napier, following the earthquake of 3 February 1931.

---

the express train from Napier was delayed after the earthquake had put a bridge out of alignment.[13] But apart from the loosening of an internal beam, the Cathedral suffered no damage.[14]

But it was a different matter 27 years later, during the 1931 Hawke's Bay earthquake. According to an eyewitness account, the Cathedral rose about 60 cm into the air and then crashed to the ground. It was later described as 'a pitiful pile of bricks'.[15] The dean, who was taking a service at the time of the earthquake, was slightly injured, while one member of the congregation died and others suffered injuries. The earthquake damage throughout the Waiapu Diocese was described as 'the biggest disaster that has yet befallen the Church in New Zealand'. The value of the Cathedral building, with the organ, windows and other furniture, was put at something over £37,500. In addition, the Deanery was destroyed by fire, while chimneys had been toppled in all vicarages from Wairoa to Ormondville, the Frasertown Church had lost its spire, and the clock tower at Te Aute College had collapsed. The Diocese records were destroyed in the earthquake, and in most cases there were no surviving plans to assist decisions regarding the state of damaged buildings.[16]

One of the many other victims of the earthquake in Napier was St Paul's Presbyterian Church. Designed by local architects J.A. Louis Hay and Walter P. Finch, it replaced an earlier wooden structure (built in 1861) which burned down in 1929. Like St John's, the new St Paul's was built of brick,[17] and although it survived the quake itself, it was gutted by the subsequent fire. This occurred only eight days before the church was due to be dedicated, so it was never used. In 1932 the image of the burned-out ruin of St Paul's became the subject of a well-known painting, *Renaissance*, by English-born artist Roland Hipkins, who had come to New Zealand a decade earlier and taught at Napier. He was one of a number of artists who promoted the development of a distinctive local style, and *Renaissance* can be seen as having wide implications during a period of artistic and cultural transition in New Zealand. The gaunt brick walls of the roofless church dominate the foreground of the picture, while a sprouting tree fern and construction sites beyond symbolise progress and the rebuilding of a shattered city. According to art historian Michael Dunn, Hipkins was determined to show the spirit of people triumphing over adversity, and in so doing he created an image that has gained national as well as regional significance.[18]

Following the 1931 Hawke's Bay earthquake a 'temporary' wooden building served as a cathedral for the Diocese of Waiapu for the next quarter of a century. In 1946 a decision was made to rebuild, and the foundation stone was laid on 12 October 1955. The building was finally completed in 1965 and the new Waiapu Anglican Cathedral of St John was consecrated two years later.[19]

THE COUNTRY'S WORST DISASTER

On the morning of 3 February 1931 New Zealand experienced its worst disaster, resulting in 261 deaths. Shortly before 11 o'clock a force 7.8 earthquake struck Hawke's Bay and was felt over nearly the whole of the southern half of the North Island and as far south as Invercargill. Napier, which then had a population of 16,025, was subjected to a violent upheaval. On the foreshore, the sea rushed away from the beach for several hundred metres, and then rolled back. Following the earthquake, fires began in chemist shops, fuelled by flammable liquids. They spread rapidly, and the business district was soon reduced to a smouldering ruin.

12
RANGIATEA

A UNIQUE BLEND:
RANGIATEA

RANGIATEA AT OTAKI stood apart from the churches of New Zealand. Although others may have claimed better examples of Maori arts and crafts, Rangiatea, whose name meant 'the Abode of the Absolute', was the first to combine Maori and European building traditions on such a grand scale. It possessed what has been described as 'a peculiar appeal to the Maori people',[1] and its design and construction were undertaken by a redoubtable triumvirate, of Ngati Toa rangatira Te Rauparaha, Bishop Octavius Hadfield and Revd Samuel Williams. Rangiatea was the oldest building of its type in the country, but in 1995, after over 140 years of service, it was destroyed by fire. In subsequent years a fine new church has risen from the ashes.

Octavius Hadfield was born on the Isle of Wight, England, in 1814. In 1837 he joined the Church Missionary Society, and the following year he sailed for Sydney. From there he crossed to Waimate North, New Zealand, where he taught at the mission school and studied the Maori language. He then volunteered for missionary work on the Kapiti Coast, north-east of Wellington. He took up his duties in November 1839, based at Waikanae and Otaki and working among the Te Ati Awa, Ngati Toa and Ngati Raukawa people.

At Waikanae Hadfield befriended Te Rauparaha, who provided timber for the building of a church. Construction was under way by October 1842 when Bishop Selwyn visited the settlement, and was completed the following year. The church established a pattern that would later be followed by Rangiatea, with three massive central posts, tukutuku panels and kowhaiwhai rafter patterns, all of which were features typical of a whare whakairo, a decorated meeting house. Although the Waikanae church was sizeable, an estimated 21.6 m long by 11 m wide, it was not the largest such structure; missionary William Colenso had visited a church at Waihou (Thames) which he declared was 29 m by 12.2 m.[2]

The Waikanae church became the first substantial building for worship on the Kapiti Coast, and with the possible exception of another at Okukari, in Queen Charlotte Sound, was the outstanding such structure in Hadfield's extensive domain until the building of Rangiatea.[3] In mid-1846 two British officers journeying overland from Wellington to

ABOVE: *Charles D. Barraud*, Interior of Rangiatea Church at Otaki, *chromolithograph, c.1852*.
PREVIOUS: *Centennial celebrations following the refurbishment of Rangiatea, Otaki, in 1950*.

Auckland reported on what they saw at Waikanae: 'The native church is well worth seeing, as an example of maori [sic] ingenuity.' They rather underestimated its size, at 12 m long by 9 m broad, but described the interior as 'neatly fitted up with reeds, and a kind of arabesque painting on the wood work, has a good effect, and is executed with some taste'.[4]

TE RAUPARAHA

Te Rauparaha was born in the 1760s, either at Kawhia or Maungataupiri. He quickly rose to leadership of Ngati Toa and, following defeat by Waikato tribes, led his people to the south of the North Island and established himself on Kapiti Island. His victories in the southern half of the North Island and the northern part of the South Island changed forever life in this part of the country. Te Rauparaha was a leader of great ability and a man of many parts, a leader as skilled in the arts of peace as in the arts of war. He was, for example, a renowned orator and diplomat, epitomising leadership in the traditional Maori style, but was regarded by early Europeans as a treacherous old-time chief. His resistance to the New Zealand Company's plans led to conflict between settlers and Maori – he was present with Te Rangihaeata at Wairau in 1843 when 22 Europeans and four Maori were killed. Later, in 1846, Governor Grey ordered him to be captured and held prisoner. When released in January 1848, Te Rauparaha went to live at Otaki, where he played a major role in the building of Rangiatea. He died late the following year.

Selecting giant totara

Inspired by the Waikanae church, Te Rauparaha determined to have a larger and more impressive structure at his own pa at Otaki. Beginning in May 1844, local tribes selected suitable trees from the dense bush on the lower slopes of the Tararua Range. Totara were felled onto a bed of branches to minimise breakages, and then stripped of limbs and floated down the Ohau and Waikawa Rivers. When the trunks reached the coast, they were lashed into rafts and towed to where they were to be prepared for the construction of the church, sited at the foot of the small hill, Mutikotiko.

Rangiatea, the name of the new church, referred to the former Polynesian homeland of Ra'iatea, about 240 km west-north-west of Tahiti. Rangiatea was also a temple in the 12th Maori heaven, the place where the whatu kura, the sacred stones connected with teaching in the highest schools of learning, were deposited on the ahurewa, or altar.[5]

Rangiatea was constructed by voluntary labour, and involved upwards of 400 Maori workers.[6] Men adzed the timbers into shape while women wove the decorative tukutuku panels. A massive 26.2 m-long tauhu, or ridgepole, was supported on three main pillars.[7] It is claimed the ridgepole was originally some 29.5 m long, but the Revd Samuel Williams, who was responsible for the overall supervision of the project, lacked faith in the Maori workers' ability to hoist such an immense piece of timber into position, so one night he took a saw to it and reduced it by 3 m.[8] The walls of the church consisted of large 10 cm-thick slabs of totara that were set up on end at equal distances of some 90 cm apart. They were coloured with kokowai (red ochre) and white paint, and the spaces in between filled with reed work, while the roof of the church was clad with split shingles.[9]

Setbacks and interruptions

The construction of Rangiatea faced several setbacks and interruptions. In late 1844 Hadfield became seriously ill and had to retire to Wellington for treatment. Until he was well enough to return he was replaced by Samuel Williams, his own brother-in-law and the son of fellow missionary Henry Williams. The building of Rangiatea had ground to a halt during Hadfield's absence, but Samuel Williams was determined to revive the project.

Another factor affecting progress was the tension following the first serious clash between Maori and British settlers, at Wairau, in what is now Marlborough. There were clashes in the Wellington region, and rumours of a planned attack on the settlement. As a result, in May 1846 Governor George Grey decided Te Rauparaha could not be trusted, and had him detained. He was allowed to return to Otaki in January 1848, whereupon Samuel Williams enlisted his support for the church-building project. The chief used his mana and influence to encourage allied tribes to provide both materials and food for the workforce. A recuperated Octavius Hadfield returned to Otaki in November 1849, by which time the inside of the church was finished, apart from the flooring, although there was still work to be done on the exterior.[10]

Death of Te Rauparaha

The now aged Te Rauparaha kept an eye on the progress of Rangiatea, obtaining the services of Te Arawa carvers to provide the adzed finish to the internal wall-panels and columns.[11] When he died on 27 November 1849 he was buried a short distance from the door of the church, although he is believed to have been reinterred on the island of Kapiti. Rangiatea was unfinished at the time of the funeral, but a Wellington newspaper noted the completion of the life of the rangatira: 'Thus terminated the earthly career of one of the most artful and successful chiefs that ever addressed a Tribe of New Zealanders, or carried terror into the ranks of his opponents; but as the grave swallows all distinctions that made us foes – the miserable past is forgotten – and all alike lie down in peace together.'[12]

In April 1848 the majority of Te Ati Awa had left Waikanae for their former home near Waitara, Taranaki. Otaki now superseded Waikanae in importance as a religious and social centre, and Rangiatea became the principal place of worship in the district. The Waikanae church fell into disrepair, threatened by regular sand-drifts. It was abandoned, and by 1851 it lay in ruins.[13]

Rangiatea was finished by 1852, when Hadfield conducted his first confirmation in the building, although at least one service had been held earlier.[14] Obviously influenced by the precedent at Waikanae, the architecture of Rangiatea was a unique blend of Maori and English church design. The ridgepole represented the belief in the one true Christian God, and the three 12 m-high central pillars symbolised the Holy Trinity. The mangopare (hammerhead shark pattern) painted on the rafters signified power and prestige, as associated with the warrior of the sea. The woven reed tukutuku panels displayed the purapura whetu (star dust), said to have been based on the patterns of the Milky Way.

Adding the ancestors later

At first Rangiatea did not have the intricate carvings of ancestor figures found in other Maori churches. It seems that the missionaries did not consider such carvings appropriate for a house of worship, but such elements were added later.[15] With the exception of its tall lancet windows, the church was entirely of Maori workmanship.[16] The beams, for example, were dove-tailed and affixed in the traditional style without nails.[17] It was very much a Maori building and a development of the traditional whare form, with a much steeper roof. Rangiatea measured 24.4 m long, 11 m wide and 12.2 m high, and because it was longer than the church itself, the central ridgepole provided projecting eaves at each end of the building.[18]

The decade that followed the completion of Rangiatea has been termed a 'golden age', when it enjoyed crowded congregations of Ngati Raukawa, Ngati Toa and other worshippers. But later came inter-tribal feuds, and by 1870, when Hadfield left Otaki to become bishop of Wellington, much of the church's glory had gone. By the early 1900s the Sunday congregation was regularly reduced to single figures.[19]

Time for renovation

As originally built, Rangiatea was clad with raupo, which was later replaced by weatherboards. In its early days the church had no external supports, but in 1884 12 buttresses were erected around the perimeter. By the early 1900s decay was discovered in the lower sections of the three great totara columns, and an extensive programme of renovation was undertaken by the Wellington Diocesan architect, F. de J. Clere; concrete foundations were put in, the roof was strengthened and walls straightened. And while the introduction of additional supports for the beams carrying the rafters did detract somewhat from the original sense of openness and airiness, it was essential for the building's survival. Following the work, New Zealand's first architectural periodical, *Progress,* recorded: 'All New Zealanders are to be congratulated upon the saving of this relic of early Maori Christian work. Nothing like it could be built again, and its loss would have been a national one.'[20]

Another major renovation of Rangiatea was carried out in the late 1940s, with practical as well as 'substantial financial assistance' provided by the Department of Maori Affairs. A major force behind the project was Sir Apirana Ngata, who had stimulated a revival of interest in Maori language, history and traditions, and encouraged the building of meeting houses. He now arranged for tukutuku experts from Ngati Porou at Ruatoria to help with the restoration of the interior of Rangiatea.

Celebrating the century

In March 1950 celebrations were held at Otaki to commemorate the restoration of Rangiatea and its century of service. A large number of visitors came from all over New Zealand, and it was the last public appearance, beyond his tribal boundaries, of Sir Apirana Ngata, who died a few weeks after the event. The service of thanksgiving and benediction was held on Saturday, 18 March 1950, and attended by the Governor-General and Lady Freyberg, the Right Revd F.A. Bennett, Bishop of Aotearoa, Bishop Owen of Wellington, clergymen from non-Anglican churches, and King Koroki.

During his sermon Bishop Bennett observed that, 'while there are other Maori churches of greater beauty, there is none of more significance to the Maori people, and certainly none as old'. He also noted that Rangiatea was 'singular' by virtue of not having been dedicated to a saint. Further, in naming the church after their Polynesian homeland of Ra'iatea, the Maori people had united their ancient mythology with Christianity.[21]

When Rangiatea first opened, Queen Victoria had presented an altar cloth worked in crochet 'by her own hands'. This kahu papura (purple cloth) was decorated with a grape motif and was one of the most venerated articles in Rangiatea, but it had since deteriorated to such a state that it could no longer be used. When Bishop Bennett visited England in 1948 to attend the Lambeth Conference, he took with him what remained of the kahu papura. The King's private secretary had the Royal Art Society make a new altar cloth. At the centennial service the new cloth was covered by the Union Jack and unveiled by Sir Bernard Freyberg. In addition, Bishop Owen dedicated a new organ for the church, and a new pulpit was carved by Te Arawa craftsmen.[22]

Rangiatea burns to the ground

At the time of its construction the government estimate of the value of Rangiatea was £2000 to £3000, and Sir George Grey declared that the church would stand for a century, at least.[23] In fact it stood for some 143 years, until the early morning of 7 October 1995 when it burned to the ground. The cause of the fire is believed to be arson and, ironically, the church had only just undergone another extensive restoration.[24]

After a period of mourning the decision was made to rebuild and to repeat the form of Rangiatea as it was at the time of the fire. Following the gifting of materials and labour, the new church rose from the ashes. The most significant change was in deference to the Revd Williams's nocturnal shortening of the ridgepole, which had eliminated the possibility of a porch in the original building. As a result, in the new Rangiatea a small side porch was extended and opened, thereby suggesting the traditional architecture of the Pacific and reaffirming the church's strong connection with the Polynesian homeland of Ra'iatea.[25]

13
PARLIAMENT BUILDINGS

WELLINGTON GOTHIC:
PARLIAMENT BUILDINGS

THE PERIOD 1906-07 was a time of significant and dramatic changes in New Zealand. In June 1906 Premier Richard Seddon died on board ship while returning home from Australia, in September 1907 New Zealand graduated from a colony to a dominion, and two and a half months later Parliament Buildings in Wellington were destroyed by a spectacular fire. It was considered the country's most historic building and, ironically, was mostly constructed of timber rather than brick in order to withstand another hazard, earthquakes, described at the time as the 'restlessness of Mother Earth'.[1]

The destruction of Parliament Buildings in the early morning of 11 December 1907 inspired some extremely imaginative reporting. The list of synonyms for 'red' was all but exhausted, and the fire was frequently compared to an insatiable beast. According to the *Evening Post*: 'At a quarter to three, just before morning rolled away the blankets of night, lurid smoke rolled and billowed upwards, like volumes of blood-red steam from an inferno. Myriads of rosy sparks and larger glowing fragments soared upwards, and then zigzagged to the ground – a shower of rosy snow.'

'The most historic pile'
The late Parliament Buildings, described by the *Auckland Star* as 'the most historic pile in the Dominion',[2] fronted Molesworth Street and were flanked by Sydney and Hill Streets. They had begun life as the Council Chamber for the Wellington Provincial Government, which was first occupied in 1858. Anticipating the removal of the seat of government from Auckland to Wellington, the Provincial Council had included chambers in its new building for both the House of Representatives and the Legislative Council. The seat of government did not in fact move south until 1865, and soon afterwards the building was enlarged to better serve its new function. It survived as such until 1872, when parts were found to be so affected by dry rot that a major reconstruction became necessary.

ABOVE: *Parliament Buildings in 1906, viewed from Molesworth Street. The statue on the left is of John Balance, Premier of New Zealand 1891–1893.* PREVIOUS: *Parliament Buildings, Wellington, viewed from Sydney Street. Parliament's Lower House and Legislative Council, on the left, were destroyed in the fire of 11 December 1907, while the Assembly Library, on the right, survived.*

The next phase in the evolution of Parliament Buildings was planned by New Zealand's first (and only) Colonial Architect, English-born William Clayton. His plan for the old Parliament Buildings was carried out in two stages, corresponding to the House of Representatives and the Legislative Council, and was completed in 1873 at a total cost of £11,448. At the time of construction it was suggested that the new Legislative Council chambers would be 'of unpretending exterior' and follow the style of its predecessors, in a style best described as 'Wellington Gothic'.³ Following completion, numerous alterations were made to the buildings, including the installation of electric fans and lighting, a system for purifying, heating and distributing the air, and improvements to the acoustics. Then in 1899 the complex received a new and extensive addition to house, among other things, the General Assembly Library.

> **'IN THE FIEND'S GRIP'**
> Under the headline 'A Brilliant Scene' the *Evening Post* reported the rapid destruction of Parliament Buildings: 'The flames gobbled up a turret, and reached out for other delicacies. The crunching and the crackling of it was like the noise of a monster smacking its lips over tit-bits. Glass crashed, blazing beams tottered down …' Soon 'the body of the building was in the fiend's grip, and a tremendous mass of flame, orange, yellow and red, made a thrilling picture against the deep blue of the sky. It was after half-past three. Dawn was breaking calmly. The east was striped with pink in great restful bands, all the more peaceful in contrast with the incarnate red of the conflagration.' [4]
>
> The great fire dominated the news of the day, and the same edition of the above newspaper carried another equally colourful account. By four o'clock the flames were 'roaring in exultation … Man and his machines struggled on against the merciless enemy, hydra-headed, which sprang up at unexpected points, and supported by its ally, the north wind, scorned the efforts of the little jets of water to stay its march.' At daybreak, 'A couple of gaunt chimneys pointed their ugliness against an ashen sky, and thin wisps of the pungent blue smoke of smouldering wood smarted the eyes of people who stood to gaze at the remnant of Parliament Buildings.' Lines of hoses 'like great snakes … coiled over the lawn, and hissed at the embers'. [5]

At the time of the 1907 fire, the Legislative Council (Upper House) consisted of a meeting chamber, rooms for the Speaker and Clerk to the council, as well as a members' reading room, rooms for messengers, a storeroom and what was termed a 'strangers'' reception room. The chamber itself was 30.8 m long and 7.6 m wide, and had on display three oil paintings of former Speakers of the council. There were plans to add to this gallery of worthies, but this had not happened. In an adjacent lobby, members of the Upper House were able to mingle with those of the Lower House (House of Representatives). The chamber of the latter had offices arranged around it in a horseshoe, and above were offices for Hansard reporters and galleries reserved for the press, the Speaker, the Legislative Council, ladies and the public.

The 1899 building housing the General Assembly Library (now known as the Parliamentary Library) was designed by Thomas Turnbull, who came to New Zealand in 1871 and for an initial period worked in the office of William Clayton. Turnbull later resigned from the library project in protest at changes and was replaced by Government Architect John Campbell. The new library occupied the site of the former Wellington Provincial Government Building, and displayed the ornate Gothic features of the timber structures it now adjoined. But it also broke with tradition, for it was built of brick and, fortunately as it would turn out, was fitted with a system of firewalls and iron firedoors where it adjoined the old wooden buildings. In addition to the library, the new building also housed Cabinet offices, committee rooms and the parliamentary refreshment rooms known as Bellamy's.[6]

What caused the fire?
The 1907 fire was probably caused by a short-circuit in the electrical wiring. The alarm was raised by the night watchman, who heard what sounded like a shower of rain. Investigating, he found flames in the office of the native interpreters, in an old part of the building due to be pulled down. The fire brigade arrived, and for a time the building's plastered walls held back the advance of the flames. But they crept upwards, spreading mainly through the ceilings and roof.

COLONIAL ARCHITECT
William Clayton's arrival in Wellington in 1869 coincided with a growing need for facilities, resulting from the Government's active programmes of immigration and public works. Clayton undertook the design of a wide range of public buildings, schools, hospitals and post offices. His most substantial project was the government departmental offices in Wellington, completed in 1876. Known as the Government Buildings, and while appearing to be constructed of masonry, it claimed to be the largest wooden building in the world, and is still the leading such structure in the southern hemisphere, at least.[7]

First to go was the Lower House: 'So far the green drapery only of the chamber had been swept up by the voracious tongues of the flames that lapped the whole of the walls, but from drapes to walls was a short step, and within an incredibly short time there were showers of splinters and pieces of burning matchwood falling in all directions. The destruction was complete … The whole chamber was a solid mass of flame.'[8]

Next the fire took hold of the Legislative Council buildings. Parliament was currently in recess, but the chamber was being used for matriculation and junior scholarship examinations, and many papers were lost. It being a fortnight before Christmas, the *Auckland Star* reported flames sweeping around the chamber's upper galleries with 'everything going as if the place was a pile of Christmas toys'.[9] Having levelled both Upper and Lower Houses, the fire then raged through the three-storey wooden frontage on Molesworth Street, and by five o'clock that, too, was gone. Smoke then began to pour from the roof of Bellamy's, and when the roof fell in it also 'joined the vast limbo of dead things'.[10]

Saving the Assembly Library's collection
The fire brigade's difficulties were compounded by their unfamiliarity with the building, and the fact that internally it was a maze and had been 'aptly compared to a rabbit warren'.[11] But if the brigade appeared defeated, they were able to claim a 'conquest'. Without their strenuous efforts it seems certain the Assembly Library would also have

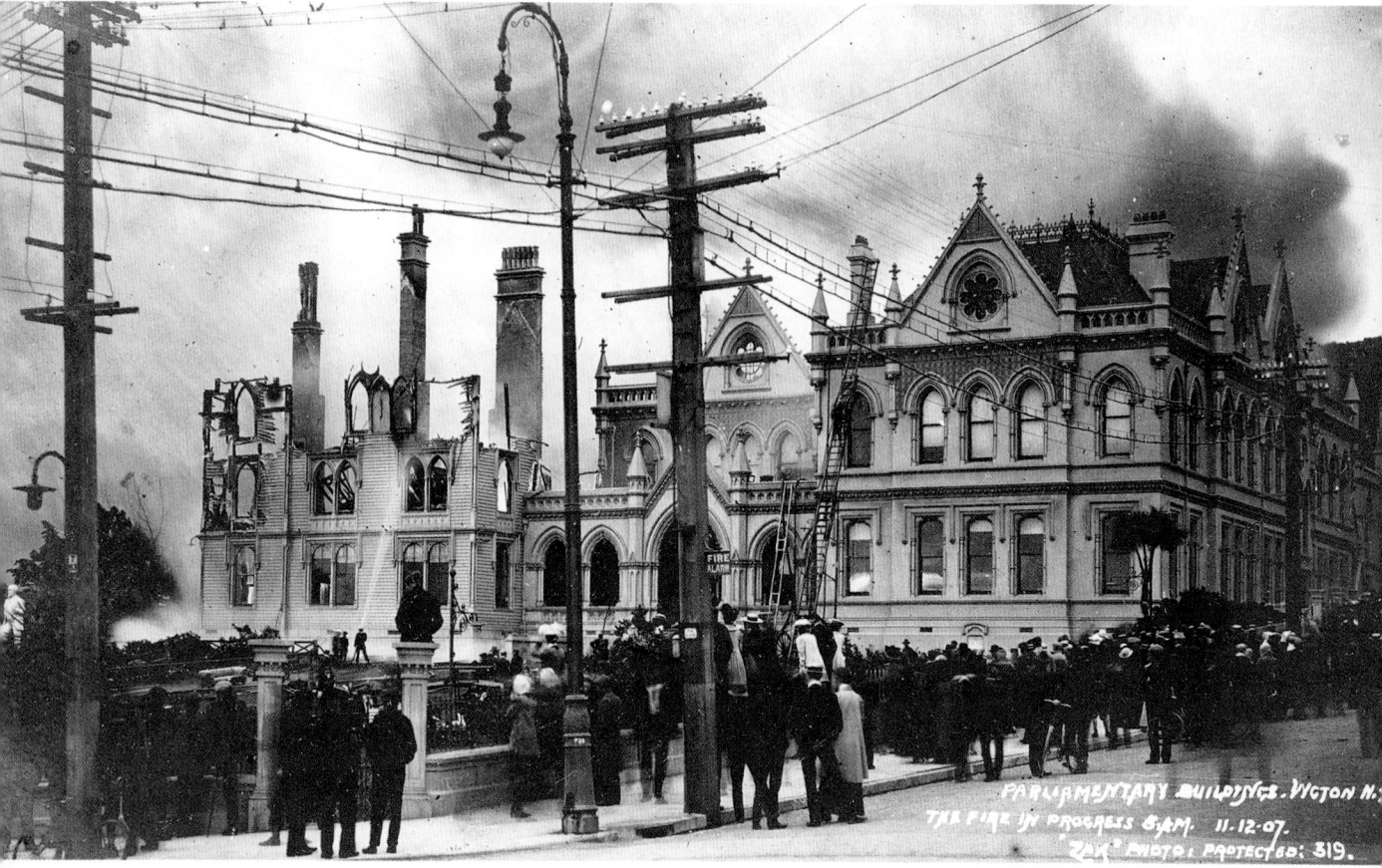

A large crowd watches as firemen attempt to deal with the blaze of 11 December 1907 that completely destroyed Parliament's Lower House and Legislative Council, on the left.

been destroyed, along with over 50,000 volumes and its invaluable collection of New Zealand literature, newspaper titles and pamphlets. Until the brigade was able to contain the advance of the fire towards his building, the Chief Librarian marshalled an army of willing onlookers to transfer the most valuable books to a safe place across the road.[12]

Meanwhile, civil servants and others associated with Parliament were engaged in salvage operations. Items of furniture, including roll-top desks, chairs, carpets and sofas, along with volumes of statutes and parliamentary papers, were piled up on the grass, while police kept an eye out for opportunistic pilferers.[13]

The press was quick to advise the public that the Government did not have a policy of insuring its buildings, for reason that it believed they were adequately equipped with fire appliances. In fact, the only such insurance the Government had was £4000 on the contents of the library.[14] Wellington's evening newspaper also took advantage of the event to remind the city of its earlier fires. Just as the Great Fire of London had led to improvements in that city, so a conflagration on Wellington's beach frontage in 1842 could take credit for the capital beginning to take its 'modern shape', by replacing its raupo whares with buildings in brick. But of the previous dozen or so major fires, the most spectacular was that which destroyed a company's chemical warehouse in March 1904; the 'evil genius of fire … gave Wellington the best pyrotechnic display that it has ever known'.[15]

Finding a temporary home for Parliament

Suggestions for a temporary home for the meeting of the next session of Parliament were quickly forthcoming, and included the Wellington Town Hall and the Canterbury Provincial Buildings in Christchurch.[16] In fact Parliament moved into nearby Government House, which the governor, Lord Plunket, vacated for a ministerial residence. The House of Representatives sat in the ballroom, while the Legislative Council occupied the conservatory.[17]

Parliament resumed on 29 June 1908, and the governor's opening speech began with a review of the previous 12 months, noting the death of the British prime minister, New Zealand's passing from the status of a colony 'to the higher plane of a Dominion', and a period of unusually dry weather that had caused problems for local farmers. His Excellency then noted the fire at Parliament Buildings, which he described as a 'threefold loss', referring to the buildings themselves and their association with the history and progress of the country.[18]

Despite suggestions Parliament should be relocated, it remained in Wellington, although it was nearly 11 years before it found a permanent home. Following a competition among New Zealand architects, a design by John Campbell was chosen for the new Parliament Building, and its foundation stone was laid in March 1912. Parliament resumed there in 1918, but the building as originally conceived was never completed on account of rising costs, earthquake risk and other factors. Instead, that part of the site would eventually be taken by the Beehive, which Government occupied in September 1979.

MOVING PARLIAMENT FROM WELLINGTON?
The need for a new home for Parliament prompted suggestions that it should not be located in Wellington. The Member for Auckland felt the seat of government should be well away from the coast, and not at a port which could be approached by an enemy ship.[19]
When a committee was appointed to report on the matter, the Member for Palmerston suggested it should investigate sites in Wairarapa, Manawatu, Nelson and Marlborough. William Massey, soon to become prime minister, considered one of the best sites in the country was the town of Frankton (Hamilton), predicting it would be 'a great centre in the future' on account of its location between Auckland and Rotorua.[20]

14

NELSON PROVINCIAL GOVERNMENT BUILDING

A HALL FROM WHERE LAWS EMANATE:
NELSON PROVINCIAL GOVERNMENT BUILDING

THE NELSON PROVINCIAL GOVERNMENT BUILDING was notable for its reference to the early 17th-century Jacobean style, which would later be employed by other architects elsewhere in New Zealand. Beginning its official duties in 1861, Nelson's building was grand on several counts; the debating chamber was claimed to be the largest in the country, while the overall cost was also huge.

But with the abolition of the provinces a decade later, alternative uses were needed. By the 1960s this once imposing building, now over 100 years old, was deteriorating, and options were running out. Finally, no use could be found for it, and in September 1969 it was demolished. The destruction of the building was controversial, but it could claim one final distinction – galvanising support for the heritage preservation movement in Nelson.

Buildings for a flourishing settlement

A decade and a half after it was founded in 1841, the New Zealand Company settlement of Nelson suddenly – in the words of architectural historian John Stacpoole – 'flourished architecturally'. It acquired several significant buildings thanks to two notable architects, William Beatson and – more especially – Maxwell Bury.

In 1850 a large public meeting was held in Nelson to discuss the anticipated benefits of provincial self-government, a system that was introduced to the colony in 1852. Nelson was one of six provinces to elect its own mini-parliament the following year, its first Superintendent being Edward Stafford, who later became Premier of New Zealand. Early meetings of the Nelson Provincial Council were held in the courthouse, described as 'an old barn with packing cases stuck on end before it', and in 1858 a committee was appointed to investigate obtaining a building of its own. A suitable design by Maxwell Bury was selected, and he was awarded a prize of £25.[1]

ABOVE: *Bird's-eye view looking west along Bridge Street c.1868. The town is dominated by the Nelson Provincial Government Building, completed some seven years earlier.* PREVIOUS: *The Nelson Provincial Government Building complex, c.1906, with its distinctive Jacobean features interpreted in timber.*

Bury's first architectural commission in Nelson had been the enlargement of Christ Church (the cathedral), also in 1858, while the same year saw the laying of the foundation stone for another of his designs, the Masonic Hall. This building was described as 'simple and elegant' and one that would reflect great credit on the architect, who had given his professional services free of charge.[2] This 'spacious and elegant edifice' was dedicated at a ceremony five months later, but the local newspaper was unable to report on the event on account of attendance being restricted to senior members of the Masonry Order.[3] The Masonic Hall was distinguished by a glass roof, which 'the colonial children … considered second only the Crystal Palace' (in London), and in its early days was used for a range of activities including public meetings, balls and concerts, and periodically as a courthouse.[4]

Friday, 26 August 1859, was a particularly big day for Maxwell Bury. Not only were the foundation stones laid for two important Nelson buildings, both designed by him, but he was responsible for supervising the associated ceremonies.[5] In addition to the

> **MAXWELL BURY**
>
> English-born Maxwell Bury trained as an engineer at an ironworks near Derby and arrived in Nelson, via Australia, in December 1854.[7] In 1863 he moved with his family to Christchurch, where he formed a brief partnership with B.W. Mountfort. He later returned to Nelson, and then shifted to Dunedin, where he was responsible for designing buildings for Otago University College in the late 1870s. He was later back in Christchurch, and then went to Sydney and England, where he died in 1912.[8]

Provincial Council Chambers there was the Nelson Institute – previously known as the Nelson Literary and Scientific Institute – which was about to get new premises in which to further its aims, one being the establishment of a museum.[6] The two buildings were sited close together, and their designs shared many common elements.

Jacobean features

The council chambers was the bigger and grander of the two, and was said to have been influenced by a Jacobean mansion, Aston Hall, near Birmingham. The Jacobean style was associated with the reign (1603–25) of King James I of England, and made much use of classical elements such as columns, pilasters and parapets, although in a more fanciful manner, along with ornamental features of the earlier Elizabethan style.

Bury's Nelson Provincial Government Building had an E-shaped floor plan similar to that of Aston Hall, while its bay windows, curved gables and ogee-roofed towers were common features of Jacobean houses. The main difference was that Bury had designed his building in wood, not stone.[9] The Jacobean style was later adopted for other prominent New Zealand buildings, including Olveston, Dunedin (1904–5), Ivey Hall, Lincoln, Canterbury (1878–80), and Holly Lea (now known as McLean's Mansion), Christchurch (1899).

Union Jack leads the way

The date (26 August 1859) had been chosen because it was Prince Albert's birthday, and was proclaimed a public holiday. Early rain cleared in time for the procession, but it resulted in muddy roads. The procession began with the Union Jack, which was followed by the Nelson Brass Band, members of the Odd Fellows and Masons and, further back, the Superintendent of the Province, John Perry Robinson (who had succeeded Edward Stafford), visiting Austrian geologist Ferdinand Hochstetter, the Bishop of Nelson, and the architect and his contractors.[10] When the procession arrived at the site of the Provincial Government Building, Bury gave a description of the ceremony about to take place, and read a list of items deposited in a glass bottle that would be placed under the foundation stone. They included recent copies of the *Nelson Examiner*, autographs of Government officials, a printed list of events at the Nelson racecourse, various mineral specimens, and samples of locally manufactured cloth.[11]

The Superintendent then gave the foundation stone the customary three blows with a mallet, and it was declared well and truly laid. After 'God Save the Queen' and prayers, he admitted to an initial lack of enthusiasm for such a formal ceremony, but declared he had been swayed by the wish of his fellow citizens. He then launched into a lengthy denunciation of central government, emphasising the innumerable benefits of the provincial system to which this building would be dedicated. He wondered if many fellow citizens had imagined, when leaving Britain, that they were coming to a country where they would have to live under an 'arbitrary government' which resulted in 'a state of misery and destitution'. Robinson claimed to have been 'attached' to the cause of local self-government for nearly 30 years, and believed the central government had done 'literally nothing' for those early settlers, who had been reduced to a state of 'misery and degradation'.

An historic spot

The new Nelson Provincial Government Building was situated on the very spot at which the early settlers had held their first political meeting nine years earlier. It would, according to Robinson, be the hall 'from whence the laws of their future local government are to emanate'. Continuing his comparison of the two systems, Robinson accused centralisation of standing for 'dogmatism' and 'the supremacy of an irresponsible oligarchy', while local self-government was distinguished by 'discussion' and 'the practical assertion of the rights and responsibilities of freemen'. In his view local government bred 'self-respect and moral dignity', whereas the other undermined such things and encouraged 'subservient sycophancy and moral degradation'. Furthermore, centralisation 'brutalizes and debases, begets and fosters a grovelling material selfishness, shuns the light of day [and] works secretly, stealthily, and by indirect and tortuous courses behind the backs of men'. Robinson completed his litany of the ills of central government by accusing it of stirring up 'jealousies, and strifes, and heartburnings' and being identified with 'the forced repression of man's natural progress'.

When the Superintendent had finally finished, the procession reformed and made its way to the nearby site set aside for the Nelson Institute. As before, Bury read a list of the contents of the time capsule, which were similar to the previous one. On this occasion lack of space prevented the inclusion of a 'very valuable' item which had been offered only that morning. Bury's identification of it as 'Mr. Mackay's celebrated umbrella' caused mirth among the crowd, being a reference to an altercation involving two Members during the sitting of New Zealand's first Parliament in 1854.[12]

Ferdinand Hochstetter had agreed to a request to lengthen his stay in the Nelson district in order to lay the foundation stone for a building dedicated to 'the noble purpose of advancing art and science'. It would 'give shelter to the treasures of literature … and

Ogee-roofed towers and elegantly curved gable ends were a feature of Maxwell Bury's design for the Nelson Provincial Government Building.

collections of objects from every department of the kingdom of Nature'. Hochstetter concluded his speech with an appropriate geological reference, hoping that the enterprise's foundations were as securely laid as the Southern Alps.[13]

Largest in New Zealand

The Provincial Government Building cost £9000 – a huge sum in those days – and hosted its first meeting of the council on 30 April 1861.[14] The chamber itself was 21.3 m long, 9.1 m wide and 6.4 m high, and was said to be the largest such room in New Zealand. It contained a reporters' and a strangers' gallery, and the interior fittings were of 'polished red pine' (rimu).[15]

Shortly, in May 1861, the Institute opened with an exhibition described as a 'humble imitation' of the Great Exhibition held in London exactly a decade earlier. Nelsonians were said to be 'remarkable for their powers of imitation on a small scale', and this collection of borrowed paintings, relics, coins, stuffed birds, geological specimens and, among other rarities, 'native curiosities', was claimed as the first of its kind in the colony.[16] Although the Institute's new building had cost considerably more than the original contracted price, it was said that 'its utility, internal arrangement, and good taste, are self-evident, and reflect great credit on both architect and builder'.[17]

Finding alternative uses

With growing support for centralised government, the effectiveness of New Zealand's provincial governments was reduced. Nelson Superintendent John Perry Robinson was drowned in January 1865, while visiting the West Coast, so did not live to see the dominance of the system against which he had spoken so vehemently. Following the abolition in 1875, Nelson's Provincial Government Building was put to a range of alternative uses.[18]

Bury's Institute building continued, but in the early morning of 25 February 1906 it was destroyed by fire. Books in its extensive library were damaged by smoke and water, and while glass display cases in a recent addition which housed the museum were smashed, their contents survived, among them 'a Maori god, said to be worth a considerable sum'. The building was of totara and said to be in good condition, while its destruction robbed Nelson of 'a characteristic landmark'.[19]

Nine years later, in July 1915, Bury's Masonic Hall was demolished. The Southern Star Lodge No. 1037 of Free and Accepted Masons had got into financial difficulty soon after their building was opened in 1858, and sold it for £750 in 1863 (its value when new was estimated at £1000). For a period the lodge rented its old premises until it was able to afford a new building of its own. The site of the 1858 Masonic Hall is now occupied by the Nelson Provincial Museum.[20]

Finally demolished

Meanwhile, the sole survivor of Bury's three buildings, the Nelson Provincial Government Building, remained at the centre of the city's civic life, and was a venue for events ranging from balls to poultry shows. But after a century of varied service, it had deteriorated and developed leaks.[21] When the Ministry of Works proposed demolishing the buildings and replacing them with a new courthouse in 1966, there was strong public reaction. The Government offered the building to the Nelson City Council, along with a grant for necessary renovations, but the council considered it insufficient and declined. Unfortunately, no use could be now found for this superb old building, and it was finally demolished in September 1969, and the time capsule retrieved.[22]

A POSITIVE OUTCOME

While the demolition in 1969 of the Nelson Provincial Government Building represented an unfortunate loss, that event has been identified as a key moment in the history of building preservation in this country. It prompted the formation of an active 'preservation lobby', consisting of the local committee of the New Zealand Historic Places Trust and others. Although such efforts were unable to save several other notable buildings, one success was the preservation of a precinct of relatively humble buildings, as opposed to a single structure.[23]

15

T.J. EDMONDS LTD FACTORY

HOME OF THE RISING SUN:
T.J. EDMONDS LTD FACTORY

A CHRISTCHURCH FACTORY was responsible for a product once found in every home in New Zealand. The Edmonds factory building was a much-admired landmark, set in park-like grounds and dominated by a celebration of the company's well-known trademark, while the business was further distinguished by the philanthropy of its founder. But the building and its grounds fell victim to changing times, within both the nation's kitchens and also the corporate sector. In 1970 Canadian Joni Mitchell sang 'They paved paradise / And put up a parking lot'; two decades later the historic Christchurch building was demolished by its new corporate owners to make way for a petrol station.

Thomas John Edmonds was born in Poplar, East London, and in 1879 he and his wife, Jane, sailed on the *Waitangi* for New Zealand. They arrived in Lyttelton on 26 September and opened a grocery store on the corner of what is now Edmonds and Randolph Streets, Woolston, Christchurch. Before long Thomas had detected the need for a reliable baking powder, so while Jane managed the shop he drew on earlier work experience with a major confectionary manufacturer in London and developed a recipe of his own.[1]

'Sure to Rise'
The story goes that when Edmonds was asked by a doubting customer whether his new powder was effective he advised it was 'sure to rise'. Thus one of New Zealand's most durable commercial slogans and trademarks was born. But the product was hardly an overnight success. Edmonds had left samples with merchants anticipating a flood of orders, but nothing happened. Realising nobody knew about his product, he began calling on householders in the region, leaving a tin and promising to take it back on his next visit if it had not proved satisfactory. As it turned out, not one tin was returned; instead, householders usually asked for more, and from that point sales began to boom.[2]

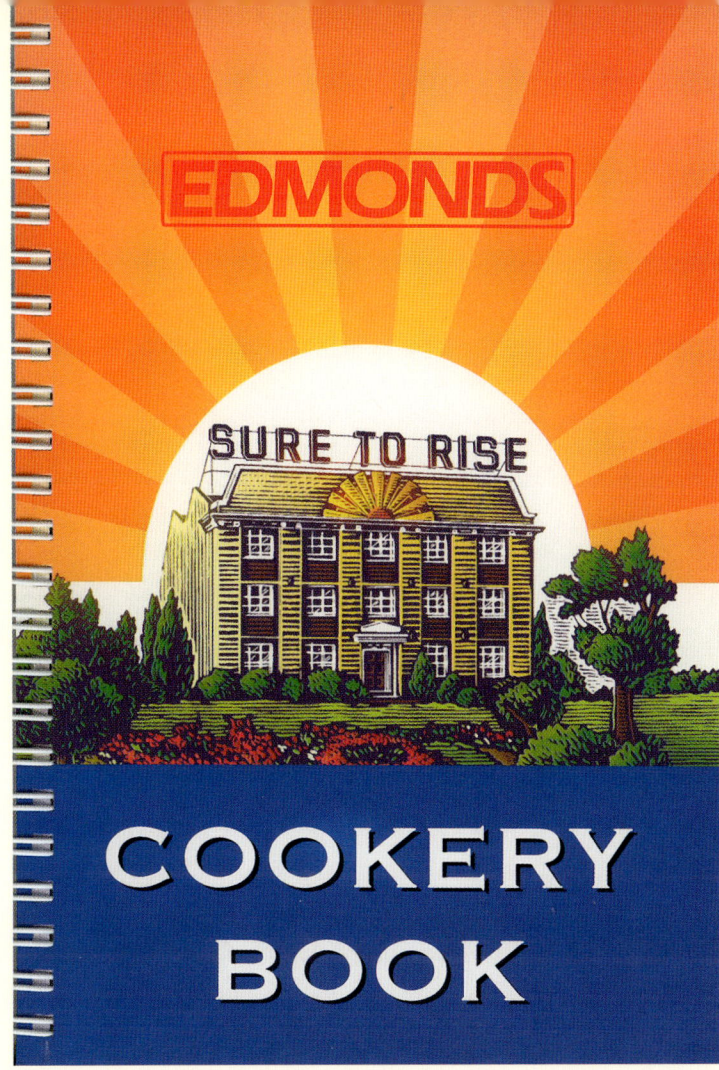

LEFT: *The Edmonds factory as it appears on the cover of the spiral-bound 49th (revised) edition of New Zealand's bestselling book.*
PREVIOUS: *Management and staff of T.J. Edmonds Ltd pose in front of their Ferry Road factory in 1940.*

We assume that once Edmonds had settled on his 'Sure to Rise' slogan, the trademark of the rising sun and plates laden with successful baking soon followed. The basic concept was hardly new; symbols of the rising sun dated from ancient civilisations and it became a popular commercial device in the 19th century, implying reliability, regularity and health-giving benefits.

As demand grew, Edmonds decided to concentrate on producing baking powder rather than operating a grocery. In the early 1890s he and his wife moved to a house on the corner of Aldwins and Ferry Roads, where he built sheds in which to manufacture his powder. There were many competing products on the market, but Edmonds could soon claim his rose above them all, including Sharland's lofty Moa brand and Hudson's even more elevated Balloon brand. Edmonds' baking powder soon gained a local reputation, and was on its way to it becoming a national product. In 1907 the number of tins sold passed the half-million mark; five years later it reached one million, and just kept rising.[3]

Staff at the T.J. Edmonds factory packing tins of the company's 'Prize' baking powder ready for despatch, in 1908.

Building the new factory

The old wooden sheds were now unable to cope, and in 1920 work began on the new factory at Ferry Road. It was three-storeyed, of brick with cement plaster facings, and a distinctive saw-tooth roof to allow the maximum amount of sunlight into what was described as a 'scientifically constructed building'.[4] The building was designed on 'the most modern lines', incorporating the latest in manufacturing methods and machinery,[5] while its main feature was its imposing façade, surmounted by the 'Sure to Rise' sign, which soon became a local landmark. Coincidentally, the sunrise became a popular decorative motif during the Art Deco period of the 1920s and 1930s.

Thomas Edmonds' interests extended beyond business to astronomy, philosophy and technology, and also plants, which led to his establishment of one of the first factory gardens in New Zealand. Such landscapes reflected the idea that workplaces should provide recreational facilities and beautiful surroundings for their workers. Under the care of a head gardener, the Edmonds Factory grounds soon became a much-admired attraction with their bright floral displays in geometrically shaped beds set into the lawns. The garden was extended as adjacent land was bought, and around 1929 a glasshouse was built to accommodate Edmonds' tropical plant collection. The grounds were redesigned in 1935, and at a later stage incorporated a marigold bed planted in the form of a sunrise.[7]

> **ARCHITECT BROTHERS**
>
> John Steele Guthrie (1883–1946) and Maurice James Guthrie (1891–1968) were one of Christchurch's most active architectural firms in the first half of the 20th century. They undertook commercial buildings, churches, hotels and schools, and introduced new ideas to the city. In addition to the T.J. Edmonds Ltd factory and office building, the pair's projects included the Christchurch Boys' High School, St George's Hospital and the Victory Memorial School.[6]

If Thomas Edmonds created a national institution with his baking powder, that product was responsible for another one. The *Edmonds' Cookery Book* first appeared in 1908, with 100 recipes promoting the use of a particular brand of baking powder. The book came to reflect the times, as its recipes responded to such social changes as the widespread introduction of cooking by electricity in the 1920s and 1930s, the increased use of gas in the 1970s, the switch to metrication, and the arrival of microwave cooking.[8] Since 1908 it has sold over three million copies, making it the bestselling New Zealand book by far.[9]

Jelly and custard, too

Edmonds' choice of the sunrise device was intended to reinforce the reliability of his product, but he was also interested in the therapeutic qualities of the sun, as reflected in his personal involvement in the Radiant Health Club.[10] And while his baking powder was 'Sure to Rise', other company products carried similar claims: custard powder was 'Sure to Please', and jelly crystals were 'Sure to Set'. The Edmonds range also included an alternative baking powder, Acto. The name might imply 'active' but it is an acronym for Australian Cream of Tartar Organisation, an apparent reference to T.J. Edmonds' interest in a factory in Paramatta, Sydney, which manufactured its cream of tartar and tartaric acid, the active ingredients that cause the raising of dough. Curiously, and despite its name, Acto does not contain cream of tartar, but a food phosphate instead.

Further to its beautiful gardens, T.J. Edmonds Ltd offered conducted tours of its factory, even publishing a helpful guide for visitors. The tour followed the raw product through the various stages, from the bulkstore to the weighing and mixing machines, followed by the filling, capping, labelling and packing of tins ready for despatch to grocers throughout Australasia and the Pacific. Visitors were informed that the custard powder room was the pleasantest place in the factory, on account of the fragrance of its

A BENEVOLENT EMPLOYER

Thomas Edmonds's staff were said to enjoy working conditions 'second to none' in the country,[11] and during the Depression his factory was the first to introduce a 40-hour five-day week. The company claimed that the high proportion of long-serving employees in the factory reflected the high standard of working conditions and the policy of management towards staff.[12] The founder also extended his generosity to the community of Christchurch, contributing significantly to its architectural history. On the occasion of his company's 50th anniversary in 1929 – by which time annual sales of baking powder tins had passed 2,500,000 – Edmonds presented the city with a band rotunda (which was converted to a restaurant in 1986) and a clock tower, while also creating a park behind his showpiece factory and gardens. Among other buildings that he substantially funded or donated to Christchurch were the Theosophical Society Building and the Radiant Hall (later the Repertory Theatre).[13]

vital essences. Elsewhere they could marvel at a machine devised by Edmonds engineers that allowed jelly crystals to be packed damp, rather than dry, in order to retain their full flavour. There was also the can-making department, turning out some 110 units per minute, while 45,000 were stored on site ready for filling. Such was the output of this department that it was suggested that cans could also be added to the list of factory products. Finally, 69,000 wooden cases were also made and branded on the premises each year.[14]

Home baking declines

According to a review of the nation's manufacturers in the late 1950s, the success of T.J. Edmonds 'blends with the history of New Zealand' and 'typifies the growth of a robust young country, and faith in its economic future'.[15] But in a sign of things to come, Thomas Edmonds' first shop was demolished in 1979 after a failed attempt to preserve it for the Ferrymead Heritage Park.[16] By now, the baking powder market was also under threat, a result of the increasing use of self-raising flour and a general decline in home baking. The asset-rich T.J. Edmonds Ltd attracted the attention of corporate raiders, and it was taken over and its head office removed to Auckland to become part of Bluebird Foods.

The Ferry Road site came under the management of Brierley Cromwell Properties Ltd, and the gardens began to decline. The 'Sure to Rise' sign was taken down in 1987, even though the Historic Places Trust had earlier awarded the factory a B classification, which meant it was worthy of permanent preservation, and considered the sign an integral part of the character of the building. The sign was to be preserved at Ferrymead, while Bluebird Foods announced its intention to continue to use an image of the factory on the cover of the cookbook 'for reasons of heritage and tradition'.[17]

The T.J. Edmonds factory appears to be raked by the rising sun in this photograph by Robin Morrison, taken in 1979.

Bulldozing the building

The Edmonds Factory now lay empty. There were plans to use it as a theatre and auditorium by neighbouring Linwood High School, but talks with Brierley came to nothing. Then in October 1990, at short notice and despite last-minute attempts to forestall it, the building was bulldozed. The levelled section was subsequently sold to Mobil for a service station. The Historic Places Trust could have sought a preservation notice, but there was concern that this might lead to the Trust being required to pay compensation to the owner, or even buy the building itself. The loss of the building was seen as another example of existing legislation preventing the Trust from being able to do its job – that of protecting the nation's heritage.[18]

Parts of the garden were destroyed along with the building, but in 1991 the Christchurch City Council acquired a major portion of the original grounds and restoration was planned. Following the Christchurch earthquakes of 2010 and 2011, the Theosophical Society Building, band rotunda and Repertory Theatre were demolished. And while the Edmonds Clock Tower sustained damage, it has been stabilised prior to repair. Meanwhile, the 'Sure to Rise' brand that started it all also carries on, now part of food-manufacturing conglomerate Goodman Fielder.

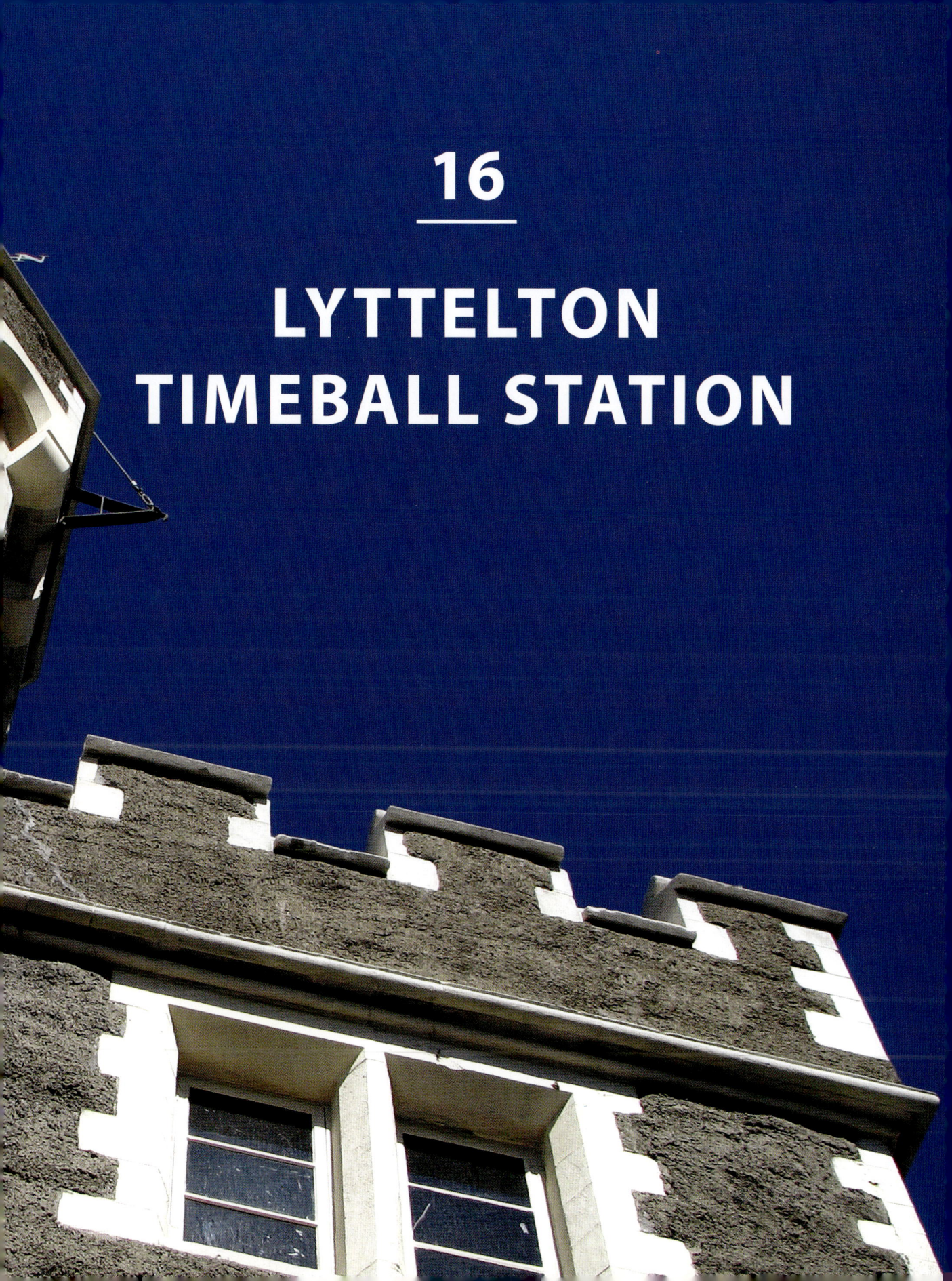

16
LYTTELTON TIMEBALL STATION

ASSISTING THOSE AT SEA: LYTTELTON TIMEBALL STATION

FOR 135 YEARS LYTTELTON'S best-known and most unusual building was a castle-like structure perched on a steep site overlooking the harbour. Its distinctiveness also extended to its original function, for it was unique in New Zealand and one of only a handful of such operating buildings around the world. Although later superseded by technology, it had been preserved as an important reminder of its contribution to local history. Unfortunately, local geology had other ideas, and the building was shaken apart and effectively destroyed by the Canterbury earthquakes of 2010 and 2011.

In the early 19th century the chronometers used during long sea voyages could accumulate errors, resulting in miscalculations of ships' positions. As a result they needed to be checked periodically against a standard (Greenwich) time. At first this was done by means of a signal from shore, leading to the introduction of the timeball in Plymouth, England, in 1829. A large ball was hoisted to the top of a pole and dropped at exactly one hour after noon. Vessels observed the moment the ball began its descent and were able to check the accuracy of their chronometers. Four years later a timeball was installed at the Greenwich Observatory, London, and the idea was soon adopted elsewhere by other maritime nations.

Sydney announced plans for an observatory with a timeball in 1847,[1] while New Zealand's first timeball was probably in Wellington, mounted above the Custom House in 1864.[2] Auckland soon had two; a timeball was dropped weekly from a mast-head at Captain Williams's residence on Smale's Point (at the waterfront end of Albert Street, cut down in the 1880s),[3] while another was installed at an elevated position above a watchmaker's business in Shortland Crescent. A local newspaper grumbled that in other centres such a useful service would be subsidised by the authorities, but in Auckland it was provided by private citizens.[4] By August 1867 Otago also had a timeball, installed at the Port Chalmers Observatory and dropped daily (Sundays excepted) at 1 p.m.[5]

ABOVE: *The Lyttelton Timeball Station with its prominent timeball – a 1.5 m zinc-covered sphere – in the lowered position.* PREVIOUS: *A corner tower of the Lyttelton Timeball Station, prior to the first Canterbury earthquake of 2010.*

FINDING POSITION AT SEA

In the early days of navigation, before radio and GPS (Global Positioning System), ships used astrolabes and sextants to find their positions at sea. These devices used celestial bodies to determine latitude, the angular distance north or south of the equator. However, longitude – the angular distance east or west of the Greenwich meridian – was not so easily measured. It was linked with the rotation of the Earth and required an accurate measure of time. Standard pendulum clocks were unreliable at sea, being affected by the rolling of the ship. After early experiments with devices using balance wheels and springs in place of a pendulum, the first true marine chronometer was designed by Yorkshire carpenter John Harrison and successfully tested in 1761–62. A copy of this was carried on Captain James Cook's second and third voyages, greatly assisting his ability to find his way around the Pacific. Another early visitor to New Zealand was naturalist Charles Darwin, who came on HMS *Beagle* in late December 1835. For this voyage Captain Robert FitzRoy packed no fewer than 22 chronometers, in order to check some of the conflicting measurements of longitude.

Canterbury leads the way

During the 1860s Canterbury enjoyed an economic boom, and Lyttelton was its main port for the shipment of locally produced wool and grain. No exact time was then kept in Canterbury, and it was felt that because it had previously pioneered the country's first public railway (in Christchurch, December 1863) the province should maintain its reputation for innovation by also having an electric timeball.[6] In 1866 the Canterbury Provincial Council considered the desirability of such a device, at an estimated cost of £500, and seven years later the necessary mechanism and machinery were imported from England. A section of land near an existing flagstaff above Lyttelton, visible from both the inner and outer habour and the heads, was provided by the Provincial Council, and cleared by prisoners from the local gaol.

Canterbury Provincial Architect Thomas Cane designed the new timeball station, described at the time as being in 'the early pointed style of architecture as applied to castellated buildings'. The walls were upwards of 0.6 m thick, of stone quarried at Sumner, laid in Portland cement and bound together with iron bands, while corner-stones, surrounds to doors and windows, cornices and battlement copings were of Oamaru stone. The timeball itself was housed in a three-storey octagonal tower, with the astronomical clock-room and a kitchen on the ground floor, two bedrooms above that, and, on the third floor, a look-out room, while access was provided by a spiral stone staircase.

The total height of the tower was 12.8 m, and above this the timeball rose another 3 m. The building was completed in 1876, whereupon the astronomical apparatus was installed with the assistance of an expert from Wellington.[7] By May 1877 the first timeball keeper had been appointed, at a salary of £50 per annum.[8]

Regular as clockwork

The timeball itself was a hollow 1.5 m-diameter sphere, made of thin zinc over a wooden frame. The associated apparatus was produced by the German firm of Siemens Bros, while the astronomical clock was from Edward Dent & Co. of London, who had made the London clock generally known as Big Ben. The ball was hoisted up to the top of a wooden (oregon) mast where it rested on a catch, which was controlled by an electromagnet operated electrically from the astronomical clock. Once released – at 1 p.m. each day (except when there was a high wind) – the ball's descent was controlled by pistons. At a later date the release of the ball was controlled by a telegraphic signal sent directly from the Wellington Observatory.[9]

By the time Lyttelton's timeball was operational there were problems with the system in Wellington. In 1877 a local citizen complained it was 'so hidden that it is comparatively useless', and suggested it be replaced by a time-gun, fired by electricity, on Mount Victoria.[10] The authorities in Auckland now provided a timeball tower, incorporated into the design of the new (1883) harbour board offices, along with an observatory and a flagstaff, all clearly visible from the harbour and some 21 m above ground level.[11] At a later point Auckland's timeball was shifted to the nearby Ferry Building, completed in 1912, although the timeball was subsequently replaced by a clock and siren.

Leaky building

The porosity of the scoria stone used in the construction of the Lyttelton Timeball Station caused it to leak, and so in 1880 it was given an application of stucco.[12] A decade later the signalman performed another public service when he observed the suspicious activities of three sailors from a ship in port. They were later charged with rolling a large boulder down the hill and 'wilfully and maliciously' damaging a Lyttelton Gas Company gasometer.[13]

In July 1891 *The Press* in Christchurch published details of the Timeball Station's operation, with some advice for those who depended on it: 'The time ball is dropped daily, Sundays and holidays excepted, at 1 o'clock p.m., New Zealand mean time, which is equivalent to 13.30 Greenwich mean time of the previous day, being calculated for the 172 degrees 30' East Longitude, and 11 hr 30 min fast of Greenwich mean time. The ball will be raised at about 2 min to 1 p.m., and will be at the top about 1 min to 1 p.m. When taking the time be careful to note the exact moment when the ball leaves the top of the mast. Should the ball not drop at the proper time, it will be kept at the top for at least two minutes, and then lowered slowly.'[14]

As a result of being narrow and having a north-south orientation, in 1868 New Zealand officially adopted a standard time (New Zealand Mean Time, or NZMT, later renamed New Zealand Standard Time, NZST) to be observed nationally, and may have been the first country do so. This standard time was ascertained at the Wellington Observatory, where it was checked by astronomical observations and against time signals received regularly from around the globe. It was then transmitted to all parts of New Zealand so that public clocks could be adjusted. The automatic time signals that went out from the standard clock at the Observatory took various forms: telegraph (sent to the General Post Office, Railways Department and South Island telegraph offices), electric lights (at the Observatory itself, to inform ships and the general public in Wellington), radio time signals (local stations ZLW and 2YA) and – of course – by timeball, at Lyttelton.[15]

> **ANTARCTIC LINKS**
>
> In addition to services to conventional shipping, the early 20th century saw Lyttelton and its Timeball Station establish links with expeditions to the Antarctic. Explorers Ernest Shackleton, Captain Robert Falcon Scott and Australian Douglas Mawson all chose Lyttelton as the New Zealand base for their operations, and set their chronometers by the local timeball prior to sailing south.[16] In anticipation of the return of the British Antarctic Expedition ship *Nimrod* to Lyttelton Harbour in 1908, flags fluttered 'Welcome' from the Timeball tower, and when the vessel came into view it was seen to have parts of her bulwarks missing, a result of 'fierce contentions with blizzards and drifting ice'.[17]

Redundancy looms

But technology was changing, and in 1910 wireless messages had been sent out each midnight from the Eiffel Tower in Paris, enabling ships to determine their longitude.[18] Timeballs became increasingly redundant with the introduction of radio time signals (which themselves have since been replaced by the satellite-based Global Positioning System, GPS), and so it was decided in 1934 that timeball signals at Lyttelton would be discontinued. The ball no longer fell but the keeper continued as a flag signalman, although in 1941 that position was also terminated.

The Timeball Station was occupied by the New Zealand Army in 1942–43, and then by various staff members of the Lyttelton harbour board until 1969. To ensure the preservation of the property, the board gifted it to the New Zealand Historic Places Trust in 1973. The Ministry of Works then began a restoration programme of both the station and the timeball mechanism, which was completed at the end of 1978.[19] And whereas flags had earlier been run up the station's flagpole to provide shipping-related information, in later years its messages were much less official. It became a popular means of conveying birthday greetings, and was even used by a local fireman to propose to his girlfriend.[20]

Earthquake damage

In 1983 the Lyttelton Timeball Station was registered as a Category 1 heritage item. The building had been thoroughly restored and was operational, and was recognised as one of only five timeball stations in working order worldwide and the only one surviving in New Zealand. But on 4 September 2010 it received partial damage in the 7.1 magnitude

Aerial view of the Lyttelton Timeball Station following damage caused by the first Christchurch earthquake, on 4 September 2010.

Christchurch earthquake. Apart from a split in the mast, the timeball mechanism itself survived relatively unscathed, and there was no sense that this would be the end of the building. Part of one side of the tower had fallen away and there were some deep cracks in the stonework, and plans were made to carry out repairs and strengthening work.

But further severe damage occurred in the 6.3 magnitude quake of 22 February 2011, and following this the decision was taken to dismantle the building. The Historic Places Trust hoped to salvage the timeball mechanism and was considering the possibility of reconstruction when a magnitude 6 earthquake struck on 13 June 2011. The tower was now toppled, and the mast and the timeball thrown down the hill.[21]

Following the collapse in June 2011, the complete building was dismantled and its fabric placed in storage. The New Zealand Historic Places Trust remained hopeful that restoration of the tower, at least, of this internationally significant structure would be possible at some future date. Those hopes received a major boost in late May 2013 with the announcement of a $1 million donation by Landmark, a charitable organisation committed to preserving this country's heritage. Historic Places Trust chief executive Bruce Chapman was confident the reconstituted Timeball Station would be a 'pretty faithful replica', and it would be built to the new earthquake-resistant building code. The project was expected to begin at the end of 2014, and would cost around $3 million.[22]

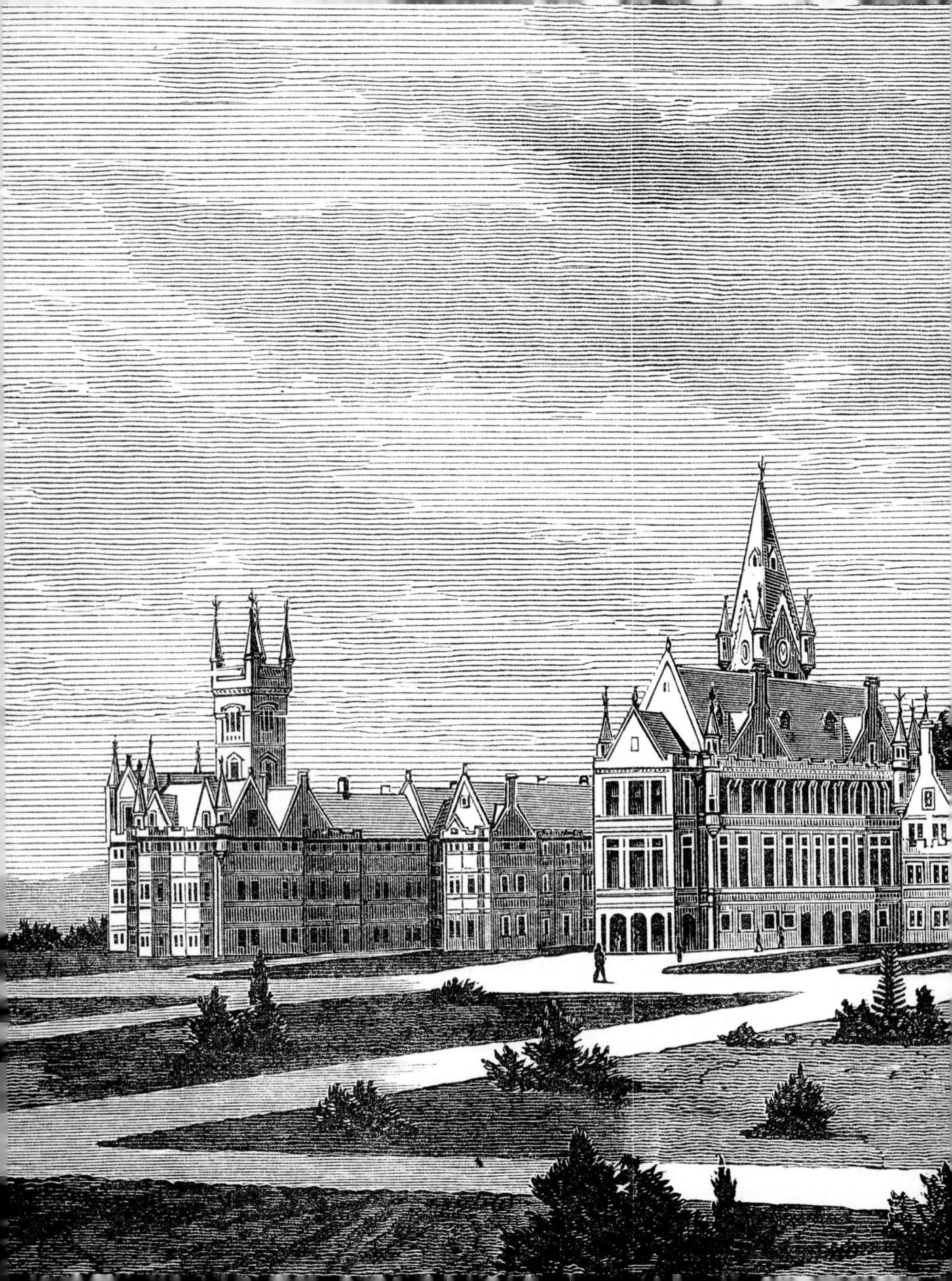

17

SEACLIFF LUNATIC ASYLUM

IN THE SCOTTISH BARONIAL MANNER:
SEACLIFF LUNATIC ASYLUM

SEACLIFF LUNATIC ASYLUM (later known as Seacliff Mental Hospital) was once the largest building in New Zealand. This elaborate and sprawling institution, with its many turrets, corbels and gabled roofs dominated by a central tower, was a massive reminder of an age with a different attitude towards the treatment of mental illness. Its architecture was exuberant, but the building was beset by problems from the time it was built until its final demolition some eight decades later. Today Seacliff is mostly remembered for two unfortunate events: an association with an acclaimed New Zealand writer, and as the site of one of this country's worst fires.

The story of the building at Seacliff begins with Dunedin's first lunatic asylum, established in 1862. It was housed in temporary quarters and intended to hold 36 patients, but by 1877 that number had swelled to 228 and necessitated a succession of extensions. The city site was now quite unsuitable, also because most of its facilities were in full view of the public. In addition, lack of space meant patients were unable to enjoy outdoor exercise and employment, then considered powerful 'curative influences' upon insanity.[1] The Government decided to relocate the asylum to a new permanent building at Seacliff, 38 km north of Dunedin. The site was a reserve of some 364 ha of sun-orientated and sloping bush-covered land, commanding an extensive view of the coastline.[2]

Pioneer asylum

The institution at Seacliff began in late 1878 with what was termed a 'pioneer asylum'.[3] This was also a temporary building and initially accommodated 14 working patients. 'Working' was the operative word, for patients were busy clearing and fencing the land and planting vegetables, and the results of this 'little agricultural experiment' were said to be well worth a visit. Patients were also engaged in the construction of the main

ABOVE: *Nursing staff of the Seacliff Lunatic Asylum pose on the lawn in front of the building, c.1890.*
PREVIOUS: *Early engraving showing the extent of the Seacliff Lunatic Asylum and its formal grounds.*

(temporary) building, completed by about mid-1879 and measuring 40.2 m by 27.4 m, using materials removed from the old asylum grounds in Dunedin.[4]

In mid-1879 the first contracts were let for Seacliff's permanent complex, designed to accommodate 500 patients.[5] The main block would have basements and foundations of concrete, a brick superstructure finished with cement and 'relieved' with Oamaru stone facings, and with stairs and steps of Port Chalmers stone. It would be flanked by wings for male and female patients, and the completed structure would be 228.6 m long. Beyond this were outhouses, including workshops, washhouses and stables, and the superintendent's residence located so as to have a commanding view of the grounds.[6]

The main block provided waiting, receiving and visiting rooms for patients, along with offices and porters' rooms, and was surmounted by a tower with a spire. The main dining hall measured 22.4 m by 13.7 m and was connected via a serving room to the kitchen and associated stores, pantries, bakehouse and sculleries. On the upper (second) floor of the central block was a massive recreation hall – 30.9 m by 13.7 m - which also served as a chapel, and was distinguished by its ceiling of 'selected colonial timbers'. The wings of the building were connected by arcades, or ambulatories, which were also linked to 'airing-courts'. Some 21.3 m long and 3.7 m wide, these were intended to provide shelter and exercise space for patients in wet weather, while recesses in the corridors were fitted with fireplaces to form a series of dayrooms. The smallest rooms for patients were each 2.7 m by 2.4 m and 'carefully ventilated', while there were hot- and cold-water baths throughout the building.[7]

Inspired by castles and châteaux
At an early stage it was suggested that Seacliff would rely more for effect on its proportions and the relationship of its parts than ornamentation, 'which would be somewhat out of character in a building of this class'. It was designed by Dunedin architect Robert Lawson, and to some extent the sloping nature of the site dictated the layout; it was three storeys high, with attics at the front, and reducing to two at the rear. Lawson was born in Scotland, and had come to New Zealand by way of Australia, so perhaps it was not surprising that the building was designed in the Scotch Baronial style.[8] Part of the 19th-century Gothic Revival, the style was inspired by medieval castles and French Renaissance châteaux and its distinguishing features included towers with small corner turrets, steep gabled roofs and battlements. Balmoral, Queen Victoria's mansion in the Scottish Highlands, was a well-known example of the style which proved popular in the dominions of the British Empire. Lawson was one of its leading champions in New Zealand, also using it for his design of Larnach Castle in Dunedin, built between 1871 and 1887. Seacliff had many circular turrets decorating the pointed gables and tower of its central block, and was described by architectural historian John Stacpoole as 'very romantic in outline in the Scottish Baronial manner'.[9]

In January 1880 the overcrowding at the Dunedin asylum was relieved with the transfer of 55 patients to the newly opened temporary building at Seacliff. This was described as a 'branch establishment … situated in the midst of the bush', and its patients were reportedly healthy and cheerful, having cleared some 2.4 ha of land for crops. But in an ominous sign of things to come, movement of the land had already caused serious damage to the building. The local ground was now recognised as being of a 'treacherous nature', and proper drainage needed to be carried out before construction of the central block of the asylum could get under way. To further reduce the risk, the site of the

The Scottish Baronial features of Seacliffe are apparent in this view from the approaching driveway, 1912.

proposed building was shifted slightly, and an opinion on the stability of the land was obtained from the Director of the Geological Survey, Dr James Hector.[10]

Happy patients

By mid-1881, when construction of the central portion of the new building at Seacliff was under way, it was reported that all the 'mischief' done by way of slips to the temporary buildings had been repaired. The patients now appeared very well cared for; in fact, it was suggested the life they were leading was 'far better than many sane people enjoy'.[11]

However, by the middle of the following year, there were further causes for concern. An inspection of the temporary accommodation revealed a worrying shortage of water in case of fire and, given the prevalence of intemperance, there was now a licensed public house at the Seacliff railway station, only a short distance from the asylum.[12] This concern with access to alcohol was highlighted in 1883, when there were 1269 persons designated as lunatics and under treatment around the country, 140 of them at Seacliff.

While the proportion of insane persons in this country's European population was lower than that in Great Britain and Victoria, Australia, it was suggested our asylums would contain fewer people were it not for government assistance with the expense. Intemperance was identified as the main reason for the admission of patients to such institutions, although it was identified as a result rather than a cause. Among other things, intemperance was attributed to solitude, nostalgia and 'religious excitement'.[13]

Dampness in the basement

In August 1885 the new asylum at Seacliff was home to 454 patients and 51 staff. It was described as being of 'most imposing architectural design', and situated in 'a picturesque and healthy locality'. But an *Otago Daily Times* reporter found much to complain about. While it contained 'some splendid rooms, lofty, spacious corridors and roomy courts', each was very much like the others, and many finer details were found wanting. Already one patient had managed to squeeze himself through an upper-storey window, falling and injuring himself severely (although 'the shock exercised mentally rather a beneficial effect'). Among other deficiencies there was a shortage of furniture, dampness in the basement, and evidence of hurried workmanship or inferior material. Nearly everywhere there was evidence of 'small things badly done'.[14]

Lawson responded to the criticism by pointing out that Seacliff had an 'outline measurement' of considerably over half a mile' (804 m) and that it was designed according to 'the most modern and approved models'. He believed there was 'not a more satisfactory building in New Zealand'.[15] But growing concerns led to a commission of inquiry, undertaken by an engineer and two architects, one of whom was B.W. Mountfort. They heard evidence of cracking and twisting in walls, and reports that one end of a building was 42.2 cm out of plumb. They were told that the site was on a formation of boulder clay likely to expand and contract in wet and dry weather; an early warning of problems had been the movement in the foundations of the temporary buildings.[16]

After hearing all the evidence the commission recommended that part of one block be taken down and rebuilt, and that monthly observations be made on site to detect ground movement.[17] It also determined that the unsatisfactory condition of the building was due in part to the lack of drainage of the site, and defects in the construction. Lawson was held responsible, and found negligent and incompetent. With his reputation damaged, and faced with an economic downturn in New Zealand, he left for Melbourne. He remained there for 10 years, before returning in 1900 to practise in Dunedin.[18]

Sir Truby arrives

The year after the inquiry, Seacliff received its third superintendent, Frederic (later Sir) Truby King. Seacliff was now New Zealand's largest and most expensive asylum, and Truby King's plans to improve the badly designed operation included the establishment of attractive planted grounds and a productive farm, an improved sewerage system, and the promotion of fresh air and exercise and a good diet. Truby King's lengthy and beneficial association came to an end in 1923, following his appointment to the newly created post of Director of Child Health.[19]

In 1906, midway through Truby King's superintendence, Seacliff was described as 'one of the most noteworthy institutions in the country'. Recent additions to the complex included a nurses' home and a further ward for female patients (both built in 1901), and the installation of electric lighting the following year. The asylum now had a total of 700 patients, and was benefiting from its own poultry farm, vegetable gardens, milking cows, and a plantation of 250 walnut trees. The laundry was operated continually – by female patients – and it is also interesting to note that while the men's convalescent ward was equipped with billiard, sitting and dining rooms, the women had similar facilities except for a sewing room in place of a billiard room.[20]

It was also recorded in 1906 that Seacliff was fitted with 'proper fire escapes and fire alarms'. These would be tested on the night of 8 December 1942 when fire broke out in a two-storey wooden ward holding female patients. It quickly became an inferno, and with their means of escape cut off and windows sealed by shutters and gratings, 39 patients died. A subsequent commission of inquiry was unable to ascertain the cause of the fire, although one theory was that earth movement may have resulted in the crossing of electrical wires. There had been other fires, including one in 1938 which had destroyed the dining room. While the cause of one of these fires was attributed to the spontaneous combustion of coal stored in the building, it was possible the other was the result of electrical malfunction, or the action of rats, or even a patient.[21]

Beginning of the end

Everything about Seacliff was big. It was the largest architectural commission in New Zealand at the time, while the building complex was visible from passing ships and over 1000 keys were needed to fit all its doors. But the unstable ground conditions which had plagued the hospital from the outset continued, and several of its wards were deemed unsafe and had to be removed. Functions were progressively moved to Cherry Farm Hospital, a short distance north near Waikouaiti, which had opened in 1952 (and itself would close in 1992 following the move towards community care).

In 1959 Seacliff's central main block was demolished, and five years later the hospital ceased taking admissions. Demolition of the main buildings was completed in the early 1970s, and the site now contains only a group of the old service buildings, which are in private ownership. The landscaped grounds and lawns form what is now the Truby King Recreation Reserve, owned by Dunedin City.[23]

SEACLIFF'S MOST FAMOUS PATIENT

In all, some 1500 patients, either judged insane or committed for voluntary treatment, lived at Seacliff. The most famous of these was writer Janet Frame, who was committed in November 1945 for six weeks, diagnosed with incipient schizophrenia. In 1948 she received treatment at Christchurch's Sunnyside Mental Hospital, and spent most of the next eight years at Seacliff and Auckland's Avondale Hospital.[22] In 1951, while at Seacliff, Frame's first book, a collection of short stories entitled *The Lagoon: Stories*, was published by Caxton Press. The following year she was awarded one of New Zealand's most prestigious literary prizes, which the author later claimed resulted in the cancellation of her planned pre-frontal lobotomy operation. In her first full-length novel, *Owls Do Cry*, published to much acclaim in 1957, many of Frame's characters were based on childhood memories of home life and her own hospital experiences.

18
THE 1865 NEW ZEALAND EXHIBITION BUILDING

HANDSOME AND APPROPRIATE: THE 1865 NEW ZEALAND EXHIBITION BUILDING

NEW ZEALAND'S FIRST international exhibition was presented in Dunedin in 1865, in a huge building whose design drew on the covered markets of London. It was the largest masonry building in the colony at the time, and unlike other exhibition buildings that followed was intended to have a second life. The original plan was for it to become a general market, following the closure of the 1865 Exhibition, but instead it served for nearly seven decades as part of Dunedin Hospital, until it was finally demolished in 1933.

From May to October 1851 just over six million people poured through the Great Exhibition in London's Hyde Park. Commonly known as the Crystal Palace Exhibition (and more formally as the Great Exhibition of the Works of Industry of all Nations), it was the first of a series of massive international displays of manufactures and arts. New Zealand was one of 89 nations and colonies represented, sending a 'miscellaneous' collection of items, mostly natural resources with commercial potential. They included minerals, timber, kauri gum, specimens of flax, leather and humpback whale oil, while there was also a selection of Maori items – among them examples of greenstone, war clubs and fish hooks – which were presented as curios rather than as artistic or cultural objects.[1]

Our own exhibition

Fourteen years after the Great Exhibition, New Zealand presented the first international exhibition of its own, in Dunedin. The province of Otago was rich as a result of the 1861–63 goldrush, and now, combined with Southland, had the largest population in the colony. (The 1886 New Zealand census – which excluded Maori – recorded the population of Otago and Southland as 56,520, followed by Auckland [48,321] and Canterbury [38,330].) The 1865 New Zealand Exhibition had its origins some three years earlier in a fund-raising event for the building of a new St Paul's Church in Dunedin. A 'fancy bazaar' was combined with an industrial exhibition of 'useful and curious articles'

LEFT: *View just inside the entrance of the 1865 New Zealand Exhibition in Dunedin, showing clocks and pianos (mostly imported) that were part of the Otago display.*

PREVIOUS: *The 1865 New Zealand Exhibition building after it had become Dunedin Hospital (in late 1865).*

from Otago, and its success led a group of influential citizens to promote the idea of a similar exhibition on a much larger scale.[2]

At an early stage it was announced that New Zealand's first such exhibition would include local products and manufactures, as well as those from other countries which might be useful to the colony's development. The Provincial Council of Otago agreed to provide a building and £4000 towards general expenses, and a Royal Commission was appointed to organise the event. Exhibits would be divided into four main sections and 40 subclasses, following the classification system adopted by another and more recent international exhibition, in London in 1862. The commissioners also drew up a lengthy list of regulations; among other things smoking was strictly prohibited and dogs were not to be admitted to the exhibition.[3]

Joseph Paxton's design for the Crystal Palace began as a doodle produced on a piece of blotting paper. Similarly, Dunedin architect William Mason, who was one of the 14 commissioners for the New Zealand Exhibition, produced an initial sketch for a suitable building whilst at a meeting. He was awarded £800 in fees for the design, and the tender price was £10,250. However, the shifting of the ancient building from its original site to a sloping section made it necessary to include a basement, which greatly increased the cost.[4]

It was a requirement that the building would have another purpose after the exhibition, and the initial thought was that it would serve as a general market. With this in mind Mason designed a square Italianate structure with a central covered courtyard. It followed an approach established by English architect Charles Fowler, responsible for London's Covent Garden and also for Hungerford Market (1833), which would have been familiar to Mason. Fowler had included corner towers to incorporate water tanks needed for a market, structures which Mason also adopted.[5]

Shingle and clay

Construction of the New Zealand Exhibition Building began in January 1864. Concrete had been considered for the basement, but when excavations revealed a bed of shingle – ideal for such a large building – the decision was made to use local bluestone. Within three weeks, 1400 loads of stone had been used. There was also a large supply of clay close at hand for the production of the half million bricks needed, while wells sunk on site provided a plentiful supply of water for general building purposes. A 30.4 m-long workshop was constructed for the carpenters and joiners employed on the project, there being no single workplace in Dunedin able to deal with such a large contract.[6]

On 17 February 1864, following the completion of the basement, the corner-stone of the building was officially laid by the Superintendent of Otago. The day was declared a public holiday, and much anxiety surrounded the occasion, it being the Province's first such opportunity to present a 'united, yet distinctive, public demonstration'. A large procession left from the Recreation Ground for the Exhibition Building site, led by mounted troopers and schools (some 735 pupils) and followed by representatives of various trades and organisations, from watermen, butchers and fishermen to carpenters, joiners and coach-builders, along with gardeners, horticulturalists, jewellers and typographers. In addition to the corner-stone was another with a cavity to contain the obligatory documents and coins, and at 114 cm x 91 cm x 60 cm it was then the largest stone that had been so cut and finished in Otago.[7]

'Boldly rusticated'

The building was 42.7 m long and 32.9 m wide and, apart from the basement, the exterior was mainly of stuccoed brick. The ground floor was 'boldly rusticated', with a central entrance tower at each front of the building facing Great King and Cumberland Streets, the former being the main entrance and including a clock tower. Each tower was flanked by five flat-arched windows, a pattern which extended around the perimeter of the building. The upper floor was relieved by two rows of windows in groups of three, the lower series being tall and with circular heads, and the upper series being smaller and square. Above the windows ran a cornice and balustrade up to a total height of 10.7 m, with a central pilaster on each of the two sides.

The entrance towers were 7.9 m square and 36.6 m high; their upper sections incorporated balustraded balconies on each of their four sides, and above was a steeply pitched roof surmounted by a flagpole. At each corner of the building was a smaller turret or miniature campanile, each being a junior and less elaborate version of the two main towers.[8]

In the interior of the building the upper floor consisted of a gallery, 4.9 m high and 7.9 m wide, which ran around all four sides, and above the open central space was a 18.3 m-square lantern-light.[9] Following a comment made in the Provincial Council, the exhibition commissioners decided to test the structural stability of their building. A four-wheeled wagon loaded with one ton of boiler-plate was 'propelled' by four men across the upper gallery and found to cause a deflection of 8.4 mm in the 7.3 m-long girders. When the load was increased to 2.95 metric tons, equivalent to the weight of 37 adults, and requiring an extra man to move the wagon, the total deflection of the girders was just 12.7 mm. Similar tests were carried out on the main floors, all of which satisfied the commissioners that the building was fully up to its intended purpose.[10]

The building was the largest such masonry structure in the colony, and by late November it was ready and awaiting the delivery of exhibits.[11] One critic considered its interior 'exceedingly pleasant – not at all fantastic, but appropriate and solid',[12] while another thought it struck the eye 'favorably' and was both 'handsome and appropriate'. However, the same writer found the two main towers 'too ponderous', giving the whole 'a clumsy appearance'.[13] A third writer described the interior as 'light and exceedingly pleasant', although it lacked colour on walls and ceiling, 'which one is apt to consider as the necessary and only decoration' for such a building.[14] Perhaps this writer had in mind the vibrant colour scheme designed by Owen Jones for the interior ironwork of the Crystal Palace: blue, white and yellow for the verticals; blue, white and red for the curving girders; and red, white and yellow for the roof bars.[15]

Preparations for the opening of the New Zealand Exhibition extended to Great King Street, a section of which had been marred with 'hollows of bog and stream' and was now levelled and metalled. Other nearby streets also received attention, so that the exhibition would be 'approachable in all directions'.[16]

NATIONS ON DISPLAY

The Great Exhibition, held at the Crystal Palace in Hyde Park, London, in 1851, was the first international exhibition of arts and manufactures. Also known as world's fairs and universal expositions, such events were mounted regularly in the main cities of Europe and the United States during the late 19th and early 20th centuries. Industrial exhibitions gave nations opportunities to display their resources provided by 'a bountiful Providence', and the benefits for world commerce were now recognised by many governments. New Zealand's first international exhibition, at Dunedin in 1865, was in fact the first in this part of the world, for Australia would not host its own such event until 1879, in Sydney.

The grand opening

Governor George Grey was unable to attend the formal opening of the exhibition as planned, so the formalities were performed by the Superintendent of the Province. At that stage the exhibition itself was not yet half-filled, due to the non-arrival of vessels with expected exhibits, and the English, Indian and French sections were all but bare. In addition to designing the building, William Mason was one of the 700 exhibitors who contributed a total of 1598 exhibits. He and his business partner, W.H. Clayton, showed a number of their drawings in the Fine Arts Section, among them of the Exhibition Building itself. Some, if not all, of these were perspective drawings by typographical painter and draughtsman George O'Brien, who frequently worked from small perspective drawings made by Mason of his proposed buildings.[17]

The exhibition was open for 102 days and closed on Saturday, 6 May 1865. At a ceremony on the last day (when there was a record 1400 visitors), the Secretary of the Commission read a report on the event, noting the 'extent, elegance, and imposing character of the building'. For the first month after opening the exhibition was disadvantaged by the non-arrival of exhibits, and the total number of visitors was 29,831 – a modest daily average of 278.

The revenue from ticket sales was less than anticipated, while finances also suffered from the non-appearance of the Governor and others from the northern part of the colony on account of the 'unsettled state of affairs'.[18] In its review the *Otago Daily Times* also noted that the past year in New Zealand had been marked by 'great disquiet', but offered a more positive outlook: 'As a mere spectacle the exhibition will melt away; as a museum it will be rapidly scattered. But the impress it leaves on Industrial Art in New Zealand will be permanent.'[19]

Finding a use for the building

When the exhibition closed it was anticipated the building would be put to its intended use as a market. However, the Provincial Council now considered such a move wasteful, and an alternative use was found. For the past decade the Dunedin Hospital had been under pressure, one of the reasons for the increasing demand for beds being the influx of men to the region following the discovery of gold. Many were single and living in boarding houses, and turned to the hospital for medical treatment. Existing facilities were inadequate, with one critic in 1862 describing the Dunedin hospital as the 'worst managed institution' he had visited in all of Australasia.

A new building was needed, and an 1863 competition for the design of a replacement was won by local architect David Ross. He later (1877) designed the Otago Museum, but his plan for the new Dunedin Hospital was not adopted. Among other things, the proposed site was problematic. It was then suggested that the 1865 Exhibition Building be converted into a hospital; the site offered room for further expansion, while the building's brick partitions made it less of a fire hazard.[20]

The estimated cost of the conversion was £2000 for the building and £2230 for the grounds. The work was not carried out by the original architect, William Mason, but by Provincial Engineer John Turnbull Thomson, who had had previous experience with hospitals as Government Surveyor in Singapore.[21] The main building was subsequently divided into eight wards, with separate quarters for staff, and offices for the storekeeper and principal surgeon. An annex, connected to the main building by an enclosed corridor, was converted into a kitchen, storeroom, waiting room for outpatients, laundry and dispensary, while a 'dead house' was built in the grounds. By late 1866 the building was ready, and was soon occupied by 104 male and 26 female patients.[22]

Gloomy and unsuitable

But there were ongoing problems with the converted premises: roofs leaked, slates blew off in high winds, and the interior was considered 'cavernous and uncomfortable'. By the early 1880s the basement accommodated the kitchen, boiler room and other facilities, and beyond that were other buildings which now included a post-mortem room and a tin shed which housed three Chinese lepers.[23] The Dunedin Hospital now developed around the main building – found increasingly gloomy and with an unsuitable layout – which was later modified to designs by Mason and Wales.[24] But by the 1930s the old main building was in a dilapidated state; its turrets were now considered unsafe and had to be removed from the towers. It was replaced by a three-storey concrete structure, which opened in 1936 and served as the main administration block of Dunedin Hospital until the late 1940s.[25]

Of the buildings designed for New Zealand's five international exhibitions (Dunedin, 1865, 1889–90 and 1925, and Christchurch, 1882 and 1906), only the first was intended to have another role after the big event was over. The 1865 New Zealand Exhibition Building was therefore unique, enjoying a second life of some 68 years as part of Dunedin Hospital until it was finally demolished in 1933.

19
INVERCARGILL POST OFFICE

THE MOST COMPLETE IN THE COUNTRY:
INVERCARGILL POST OFFICE

POST OFFICES ONCE SAT at the heart of New Zealand's towns and cities. In addition to providing vital communication services, they were often expressions of civic pride and were frequently distinguished by a prominent clock tower. But by the mid-20th century, time had moved on, and many of these buildings were considered outdated and in need of replacement. Some were able to find alternative purposes, but others, like Invercargill's magnificent Chief Post Office in Dee Street, were demolished.

New Zealand's first official postmaster was appointed in 1840 but was fairly quickly dismissed for reasons of dishonesty and drunkenness. That aside, the early post office in the colony faced various other challenges, a result of the spread of isolated settlements and the general difficulties of communication. But by 1845 there was a total of eight post offices, two of which – Nelson and Akaroa – were in the South Island, and these were followed by Port Chalmers and Dunedin (1848) and Lyttelton (1850).

Soon legislation stimulated a growth in numbers, and the first post office was established in Invercargill in 1857, having been relocated from Bluff.[1] Prior to the appointment of its first chief postmaster in 1861, the town's earliest postal duties were carried out by the local Collector of Customs.[2] Mails collected from throughout Southland were sent on from Invercargill, and a special feature of the local service, instituted in 1879, was a postal van attached to the express train to Christchurch which allowed mail to be sorted en route.[3]

A post office for Invercargill

Early Invercargill was not short of public clocks, and in 1867 another one was installed above the existing telegraph office. This would enable citizens to keep abreast of New Zealand Mean Time when it was introduced the following year. A new post office was planned, and in 1892 agreement was reached with the Government that it should

ABOVE: *above: The band rotunda has been removed and work is underway on foundations for Invercargill's new post office, to replace the 1893 building behind. This photograph was taken on 8 December 1937.*
PREVIOUS: *A line of horse-drawn cabs awaits customers in front of the Invercargill Post and Telegraph Office and band rotunda, c.1910.*

incorporate a clock tower. The Government agreed to supply the clock, while the local council would provide gas for lighting its faces, and also pay for suitable chimes.[4]

The new post office, in Dee Street, was opened on 7 August 1893. Some 900 people attended the official occasion, at which Postmaster-General Joseph Ward was guest of honour. He hailed the event as marking 'another epoch' in the history of Invercargill and district, illustrating the huge growth in local postal business by pointing out that in the past 20 years annual revenue had increased by over sevenfold, from £2368/14/0 to £17,676/13/10. Over the same period the number of letters posted from the Invercargill office had skyrocketed from 181,248 to 2,791,256, and Ward felt there could be no better indication of the growth of business throughout the colony, and its general prosperity.[5]

Thanks to the amount of shipping using nearby Bluff harbour, a considerable proportion of New Zealand's mail – either for or from Britain – was sorted at Invercargill. In fact, Ward informed the crowd that the new post office was primarily to provide additional space for that very duty. He also noted that the building was an example of where control of the design had been successfully removed from central government. In this case it had been the responsibility of the local Public Works Department engineer, and the architect William Sharp, whose 'excellent design has been carried to so successful an issue'.[6]

POSTMASTER-GENERAL AND PRIME MINISTER

Joseph Ward (1856–1930) was born in Melbourne and came across to Bluff at a young age. He left school to be a messenger with the Post and Telegraph Department, where he learned Morse code. Years later he put this experience to good use when, after giving a speech, he was unable to find anyone with the expertise to wire his notes through to the *Dominion* newspaper in Wellington – So he sat down and tapped it out himself. Ward entered politics in 1890, becoming postmaster-general the following year and prime minister for the first time in 1906.

In 1901 Ward was knighted in recognition of his work for the recently instituted universal penny postage. New Zealand was the first country to adopt the system, something Ward had been working on for 20 years. Thanks to this innovation, New Zealanders were now able to send letters for one penny to the United Kingdom and some 70 other countries and British colonies. The inauguration of the penny postal system on 1 January 1901 was a great success, resulting in an immediate increase in the amount of mail posted in New Zealand. No doubt letter writers poured into the Invercargill Post Office for their penny stamps. Five months later a large crowd gathered around the band rotunda outside when the mayor welcomed home local men who had served in the war in South Africa.[7]

WILLIAM SHARP

Born in Yorkshire in 1847, William Sharp worked for the North Eastern Railway at York, and then as assistant engineer and draughtsman on the Imperial Government Railway in Japan. He had also carried out railway surveys in North Wales, and after emigrating to New Zealand in 1878 he was appointed assistant engineer in the Public Works Department at Invercargill.[8] During the early 1900s a young architect named John Mair received his training with Sharp. He went on to become government architect in 1923 and among his numerous designs was a replacement for Sharp's Invercargill Post Office.

The contract price for the new post office was £4000. Construction proceeded according to plan, but at the time of the official opening the central tower was lacking its intended clock. A suitable timepiece had been sourced in Wellington, for £685, but it would be several months before it arrived. When this was attended to, Joseph Ward suggested Invercargill's postal building would be the most 'complete' in the country.[9] But when the new clock was installed, not everyone was impressed. Although it was centred at a height of 25.9 m in the 27.4 m tower, there were complaints it was not visible from some of the main streets and, further, that its chimes were indistinct. There were repeated suggestions by the council to the Government that the tower be raised, but nothing happened.[10]

A grand brick structure

The new post office was located between Colonial Architect William Clayton's handsome General Government Building (1876) and the Athenaeum building. It was a grand two-storey brick structure, with distinctive corner details and arched windows. Symmetrical wings flanked the central tower with its columned entrance porch, and the low roofs were crowned by masses of chimneys. The steeply pitched square tower was in French Renaissance Victorian style; it had finials in its corners and above each of the four clock faces.

The building was set back some distance from Dee Street. The large space in front was made into a central public square, and included a feature typical of the late 19th-century cityscape, a band rotunda. Citizens had first suggested such a facility back in 1885, but council decided that the playing of band music would only cause a nuisance and encourage larrikinism. However, supporters of the amenity would not be deterred and managed to raise the necessary funds. The structure was designed by local architect F.W. Burwell and opened on 15 February 1893, six months prior to the post office. In addition to brass bands, the rotunda provided a platform for a wide range of individuals to air their views, and bring further activity to the square.[11]

An early 1900s view of a lone horse-drawn cab and tram in front of the Invercargill Post and Telegraph Office and band rotunda, flanked by the Colonial Chambers (far left) and Athenaeum building (right).

A replacement post office

By the 1930s Sharp's post office was in need of replacement. It would be superseded by a new three-storey building, designed by Government Architect J.T. Mair[12] and occupying the square in front of its predecessor. Since the late 1920s the square had been home to statues of Lords Kitchener and Jellicoe and also Sir Joseph Ward, the work of Auckland sculptor William H. Feldon. Now they – and the band rotunda – had to be removed.

It was suggested that the statue of Ward be placed in the vestibule of the new post office, but the authorities could find no room for a former postmaster-general. Instead, he and the other two were re-erected in the Gala Street Reserve, although in 1970 the statue of Ward was transferred to the Borough Council in his boyhood town of Bluff.[13] Unfortunately, when the rotunda was dismantled in early 1937 its timbers were found to be in such a condition that the structure could not be re-erected.[14]

Mair's post office was officially opened on 28 July 1941, by which time Sharp's 1893 building had – apart from its tower – been demolished. The clock was still operating, although it was somewhat obscured by the new building. Three years later, following concern that it was an earthquake risk, the tower was also removed, and the clock and mechanism went into storage. In 1950 the council considered re-erecting the clock, but nothing came of the proposal, and two years later an offer by the Dunedin City Council to buy it and the chimes was declined.[15] The clock remained in storage until 1989 when, finally, it was restored using parts from the Bluff town clock, and re-installed in a new purpose-built tower in Invercargill's Wachner Place.[16]

Facing new challenges

Sharp's post office was probably done away with on account of it being both Victorian – and therefore outmoded – and a brick building in a country prone to earthquakes. But post offices in general were now facing challenges of a different sort. In 1987 the Post Office as a government department was replaced by the three independent State-owned corporations: NZ Post, Telecom and PostBank.

By means of mail and telegraph, post offices had once provided a community's contact with the outside world, but that was also changing in the new digital age. As a result, the latter half of the 20th century saw many of the nation's post offices having to find alternative uses or face oblivion. New Plymouth's Chief Post Office, for example, opened in 1907 and was demolished in 1969, although a stand-alone replica of its once prominent clock tower was built in 1985. Auckland's Imperial Baroque Central Post Office (opened in 1912) was converted into the Britomart Transport Centre, in 2003, while the 1910 Hastings Post Office, on the corner of Queen Street East and Russell Street North, was rebuilt after the 1931 earthquake and became a medical centre in 2001.[17] Equally adaptable was the 1938 Devonport Post Office, on Auckland's North Shore. In 1991 it became a private museum and visitor centre, and was subsequently converted into a shopping arcade.[18]

The loss of William Sharp's grand Invercargill Post Office was compounded by the obliteration of the much-used public space for which it provided a spectacular backdrop. The building's 1941 replacement has survived, but it was required to adjust, becoming Quest Invercargill serviced apartments. But Sharp will at least be remembered by one other significant landmark, his Invercargill Water Tower, completed in 1889. Distinguished by its polychromed brickwork, this 31.5 m structure has been likened to a 'magic Victorian mushroom'. It was restored in 1979, and has been identified by the New Zealand Historic Places Trust as one of this country's outstanding industrial monuments.[19]

20
DEE STREET HOSPITAL

A GREAT SANITARY AGENT:
DEE STREET HOSPITAL

FROM THE EARLY 1860s Southlanders were served by a hospital that grew in stages, its most imposing addition being funded by loyal citizens in honour of Queen Victoria's Diamond Jubilee. By the second decade of the 20th century the complex was proving inadequate, but it struggled on. Demolition began in the 1980s, and the majority of the buildings were removed. The property is now recognised as the site of New Zealand's oldest hospital buildings.[1]

In January 1853 New Zealand established the first of its Provincial Councils, in the main centres of Auckland, New Plymouth, Wellington, Nelson, Canterbury and Otago. Southland followed eight years later, splitting off from Otago (although reuniting in 1870), and one of its first acts was to recognise the need for a district hospital. Invercargill, Southland's leading settlement, was founded in 1856, and in its early years a hospital was needed mainly to cope with the victims of accidents. While the southern climate wasn't always agreeable, it was considered healthy and believed to be the reason for the general lack of epidemics and diseases. In fact, the district's strong breezes were described as a 'great sanitary agent'. But with the discovery of gold in Otago in 1861, the subsequent rush to Lake Wakatipu resulted in an increase in Southland's population and the incidence of sickness, much of which was attributed to the fouling of wells from household drainage.[2]

Early hospital buildings

The first hospital in Invercargill, on a reserve in Dee Street, was a short-lived and inadequate timber structure. It was replaced in 1863 by what became known as the North Wing, one of the few brick buildings in the town at the time. Three years later it was joined by the Porter's Lodge, designed by a now unknown architect and located beside the main gates to the reserve. At this stage the hospital could accommodate some two dozen patients and had a staff of six: a resident surgeon, steward, wardsman and his assistant, cook and a female domestic servant. Thus, the direct care of patients was entrusted to an all-male staff.[3]

ABOVE: *The main entrance to Invercargill's Dee Street Hospital, with the Victoria Wing (completed in 1901) on the right.* PREVIOUS: *The driveway sweeps through well-planted grounds to Dee Street Hospital's recently completed Victoria Wing in this postcard view from the early 1900s.*

The two-storey main building housed the male ward (10 beds) on the ground floor, and the female ward (10 beds) above. Beyond, a separate fever ward had an ingenious double roof of canvas below the standard iron in order to provide a continuous current of air and control the temperature of the room. Another structure housed mentally ill patients, and its melancholy function was indicated by its 'grim-looking locks, dismal bolts and ponderous window bars'. One such patient was also housed in one of the two apartments in the Porter's Lodge; the other was occupied by a nurse, the hospital now having female staff. There was also a well-equipped surgery, with drugs ranging from brimstone and treacle to croton oil — a yellowish-brown extract from an Asian plant used as a drastic purgative — and a good selection of knives and 'appliances' to deal with amputations.

Equally unappealing was the outward appearance of the building, described in September 1866 as 'a sombre-looking edifice, an antediluvian conglomeration of architectural stupidity'. This was improved a few years later with the application of Portland cement to the brickwork, and the addition of a 'handsome' verandah along the full length of the front of the main buildings.[4]

The incidence of local accidents increased during construction of the Invercargill–Mataura section of the South Island main trunk railway, and by 1873 male patients heavily outnumbered females. The constant need for more space led in 1876 to the construction of the Central Block, designed by Scottish-born Frederick W. Burwell. Described as Invercargill's earliest and probably most influential architect, he was responsible for much of the city's commercial heart.[5] But Burwell's Central Block proved to be a temporary solution only. The congestion at the Dee Street Hospital soon became 'deplorable', and was not helped by patients from Riverton who chose to not take advantage of their own perfectly adequate hospital. But some relief came in 1879 with the construction of another addition, the South Wing, also designed by Burwell.

NEW ZEALAND'S OLDEST HOSPITALS

This country's first hospitals were four colonial institutions, designed by architect Frederick Thatcher and built in Auckland, Wellington, Whanganui and New Plymouth. Of these, only the New Plymouth building (opened in 1848) has survived, but by 1903 it had ceased to function as a hospital. The following year it was sold and relocated to what is now Brooklands Park, in the city, and also acquired its present name, The Gables.[6] Invercargill's Dee Street Hospital operated for some 117 years, from around 1862 until 1979, and there are no existing hospital buildings of a comparable age in New Zealand. In fact, Invercargill is the only city that retains hospital buildings from all eras of its history. The Victoria Wing has gone, but in the view of the New Zealand Historic Places Trust the surviving Central Block and South Wing are 'unique buildings nationally' and 'have no parallel in the country'.[7]

Invercargill's heritage legacy

In 1876 it was anticipated that another of Burwell's buildings would make the architectural appearance of Invercargill's Dee Street 'famous amongst New Zealand towns'. His design, for a two-storey block with a three-storey corner section, would be finished with square towers, campanile roofs and lunettes, along with pilasters, columns and balustrades. At the same time Burwell was also the architect for Invercargill's new Masonic Hall, which would include a lodge-room illuminated and ventilated by a circular window in the ceiling, and an ornamental front of Roman Doric pilasters supporting an entablature enriched with medallions.[8] The South Wing and Central Block of the Dee Street Hospital also demonstrated Burwell's familiarity with the Classical idiom, and while neither may be a remarkable example, each has been described as being carefully composed and detailed and making a great contribution to Invercargill's legacy of built heritage.[9]

In 1880 Burwell's work was recognised when he was elected a Fellow of the Royal Institute of British Architects. But the Long Depression, which lasted from 1877 to the early 1890s, forced him to move to Australia to practise, and he died in Melbourne in 1915.[10]

A GREAT SANITARY AGENT: DEE STREET HOSPITAL

Nursing staff pose on the shaded balconies and verandah of the Victoria Wing, the most recently completed addition to the Dee Street Hospital in 1909.

Changing times

The complex of buildings that was the Dee Street Hospital continued to undergo constant change. By 1885 the fever ward had become so 'malodorous and ... decayed' that it was pulled down, and its timbers burned in case they were infested with germs. Three female wards were added in 1896, but the shortage of space continued. However, the approaching Diamond Jubilee seemed an opportune time to seek public funds for a new building. An appeal based on loyalty to both the throne and long-serving Queen Victoria herself was seen as likely to succeed – as indeed it did. Asked to subscribe £1500, the public exceeded expectations by providing £2060. Younger citizens were also able to contribute, and a subscription for children's shillings and sixpences added another £80 to the fund.

Colonial Treasurer Joseph Ward had earlier promised to use his influence to obtain a Government subsidy of £1 10s on every £1 collected locally, as had been done with a new hospital at Whanganui.[11] And although the actual Government subsidy was somewhat less, at 24/- for every £1 collected, the grand total was still some £4000 and enough to get the project under way. In June 1898, and reportedly with 'great éclat', the Governor, Lord Ranfurly, laid the foundation stone, which recorded that funding for the new building was 'subscribed by the people of Southland in commemoration of Queen Victoria's record reign'.[12]

The Victoria Wing provided two male wards with accommodation for 36 patients and was designed by the Invercargill partnership of McKenzie and Wilson.[13] It had a 'strong street presence', and became the best known of the hospital's buildings.[14] A curved drive, flanked by shrubs and ornamental urns, led to the main entrance, set between a pair of octagonal towers with turreted roofs and an open balcony above. Other works considered necessary for an institution of its 'magnitude and position' included a massive concrete and iron railing boundary fence, 194 m in length, along with other improvements to the outbuildings and grounds.

The hospital board's greatest achievement was the fact that all this work was paid for by its own natural resources. A deposit of high-quality fine gravel proved ideal for the hospital's concreting requirements, and was also sold to contractors. Another cost-saving measure was the use of prison labour, under the watchful eye of the warder. As a result, the institution now claimed 'quite an imposing appearance', especially when the town was approached by road or rail from the north. Further, its internal arrangements were said to be equally impressive, providing shelter and aid for the district's 'sick and indigent'.[15]

The battle against germs

By the early 1900s the Dee Street Hospital consisted of three main blocks of two-storey buildings with verandahs, built on the 'pavilion plan' and connected by corridors. In the never-ending battle against germs, the wards were disinfected and the floors polished with wax. But concern at the lack of cross-ventilation in 1904 led to suggestions that some of the windows in the men's ward be made to open, but this would not be possible in the women's ward on account of the severity of the southerly winds. In fact, it was claimed that in Southland 'there was ventilation which would kill more patients than foul air if it were allowed free course'.[16]

Subsequent developments at Dee Street included the completion of a new surgical block in 1910, and the adaption of the old fever ward to accommodate patients suffering from tuberculosis. But in 1917 the Inspector-General of Hospitals made the first suggestion that the site be abandoned, it being considered too small for future developments. The plan was for the Dee Street Hospital to be scrapped and superseded by a new general hospital at Kew, some 3 km south from the centre of Invercargill. This new site had earlier been used for a fever ward, and in 1918 when wooden buildings were erected here to accommodate some 50 diphtheria patients, local victims of an epidemic that resulted in 240 deaths throughout the country.

Shortly, with increasing pressure on facilities at Dee Street caused by the number of returned servicemen requiring treatment, the Southland hospital board decided to buy further land at Kew, and began planning a general hospital that could handle up to 500 patients and would entirely replace Dee Street. But several factors, including the increased cost of materials and the changes to plans required by new regulations following the 1931 Hawke's Bay earthquake, delayed construction of the new building. Finally, tenders for the main block at Kew were let in 1933, and the foundation stone was laid by the Duke of Gloucester in January 1935.[17]

Beginning of the end

Dee Street remained in use until the completion of the new hospital at Kew in 1937, and thereafter was retained for maternity services. It was upgraded in 1942 and renamed the Queen Victoria Hospital, but the new identity did not catch on with the public, and so by the late 1960s it had reverted to the Dee Street Maternity Hospital. Maternity services were transferred to Kew in 1979, and the Dee Street Hospital finally closed. Unable to find a user for the buildings, the hospital board decided to demolish all those that had not been classified by the New Zealand Historic Places Trust, as well as the Victoria Wing (which had been given a C classification).

Demolition began in September 1985, sparing Burwell's two 1870s buildings. The earlier Central Block was now used for storage by the Southland Museum and Art Gallery, while the South Wing was renamed in honour of its architect. The old resident surgeon's quarters within what is now Burwell House have been converted into accommodation for the William Hodges Fellowship, which recognises the first recorded European artist in residence in Southland who was part of Captain Cook's second voyage to New Zealand in 1773. Another survivor from the original hospital is the picturesque Porter's Lodge, built in 1866 and now reputedly Invercargill's oldest house.[18]

In 1990 the Dee Street site became home to a McDonald's restaurant and its carpark. It would not be the last time a building of historical significance was replaced by a fast food outlet; in 1992, the Hawera home of novelist and short-story writer Ronald Hugh Morrieson was demolished to make way for a Kentucky Fried Chicken outlet.

REFERENCES

INTRODUCTION

1. *New Zealand Herald*, 26 September 1873.
2. *New Zealand Parliamentary Debates*, vol. 299, 1953, pp. 722–23, 729.
3. *New Zealand Parliamentary Debates*, vol. 297, 1952, p. 507; A.H. McLintock (editor), *An Encyclopaedia of New Zealand*, Government Printer, Wellington, 1966, vol. 2, p. 664.
4. http://www.fitzroygardens.com/Cook_Index_Page.htm
5. Philip Davies, *Lost London 1870–1945*, Transatlantic Press, Hertfordshire, UK, 2009, pp. 291, 355.
6. Graeme Davison and Chris McConville (editors), *A Heritage Handbook*, Allen & Unwin, NSW, Australia, 1991, pp. 14–19.
7. *New Zealand Herald*, 26 July 1951.
8. Ross Fraser, 'Victoria Arcade', *Art New Zealand*, 20, 1981.
9. Alan Perrott, 'Monumental losses', *New Zealand Herald*, 8 October 2009.
10. *Auckland Star*, 27 August 1928.
11. Perrott.
12. 'Old Buildings in City Disappear', *New Zealand Herald*, 27 June 1964.
13. Peter Shaw, *New Zealand Architecture: From Polynesian Beginnings to 1990*, Hodder & Stoughton, Auckland, 1991, p. 190.
14. *New Zealand Herald*, 27 November 2012, B5.
15. http://www.stuff.co.nz/the-press/news/christchurchearthquake-2011/5996494/Heritagetrust-recommendations-ignored
16. http://www.chch.catholic.org.nz/?sid=2720
17. http://nz.news.yahoo.com/a/-/topstories/16539457/church-reveals-newdesigns-for-christchurch-cathedral/
18. *The Press*, 19 August 2011.
19. 'Earthquake-prone Buildings', http://www.wellington.govt.nz/services/earthquake/regs/html
20. *Weekend Herald*, 8 December 2012.
21. Maurice Shadbolt, 'The People Before' in *Figures in Light: Selected Stories*, Hodder & Stoughton, London, 1978, p. 58.
22. Terence E.R. Hodgson, *Fire & Decay: The Destruction of the Large New Zealand House*, Alister Taylor, Martinborough, 1978.
23. *New Zealand Herald*, 28 November 2012.
24. *Weekend Herald*, 25 May 2013. 'Walking in the City' exhibition at Anna Miles Gallery, High Street, Auckland.

CHAPTER 1

1. http://www.doc.govt.nz/conservation/historic/by-region/ northland/whangarei/ruapekapeka.
2. Nigel Prickett, *Landscapes of Conflict: A Field Guide to the New Zealand Wars*, Random House New Zealand, 2002, pp. 38–45.
3. *New Zealander*, 6 December 1845.
4. James Cowan, *The New Zealand Wars*, Government Printer, Wellington, 1983, vol. 1, p. 79.
5. *New Zealander*, 13 December 1845.
6. Cowan, pp. 77, 79.
7. Cowan, pp. 74–75; http://www.nzhistory.net.nz/war/northern-war/ruapekapeka
8. Cowan, pp. 75, 84.
9. Cowan, pp. 79–80.
10. Cowan, pp. 80–82.
11. Cowan, pp. 84–85.
12. Prickett, p. 46.
13. http://www.nzhistory.net.nz/war/northern-war/ruapekapeka
14. Cowan, p. 86.
15. *New Zealander*, 24 January 1846.
16. http://paperspast.natlib.govt.nz/cgi-bin/paperspast?a=d&cl=CL1.NZ&essay=1&e=
17. *New Zealander*, 24 January 1846.
18. Cowan, pp. 77, 79.
19. Prickett, p. 46.

CHAPTER 2

1. Denis Fairfax. 'Wiseman, William Saltonstall – Biography', from the *Dictionary of New Zealand Biography. Te Ara – the Encyclopedia of New Zealand*, updated 4 July2012; Daily Southern Cross, 8 December 1866.
2. *Auckland Star*, 1 September 1873.
3. *New Zealand Herald*, 31 July 1879; *Auckland Star*, 13 November 1880.
4. *New Zealand Herald*, 11 June 1881.
5. *Auckland Star*, 21 February 1898.
6. *Auckland Star*, 5 November 1897.
7. *New Zealand Parliamentary Debates*, 2 November 1898, vol. 105, p. 619; *Auckland Star*, 1 April 1901.
8. http://www.historic.org.nz/corporate/registersearch/ProfessionalBio/Professional.aspx?CPName=Arnold%2C+Charles
9. *Auckland Star*, 26 October 1903.
10. Terence E.R. Hodgson, *Fire & Decay: The Destruction of the Large New Zealand House*, Alister Taylor, Martinborough, 1978, pp. 91–92.
11. *Observer*, 6 July 1901.
12. *Auckland Star*, 25 June & 4 July 1902, 13 March 1903.
13. *Auckland Star*, 25 February 1903.
14. *Auckland Star*, 2 March 1903.
15. *Observer*, 7 March 1903.
16. *Auckland Star*, 24 October 1903.
17. *Auckland Star*, 18 March 1903.
18. *Auckland Star*, 26 October 1903.
19. *Auckland Star*, 26 October 1903.
20. *Auckland Star*, 26 October 1903.
21. *Auckland Star*, 22 December 1903.
22. *Auckland Star*, 20 August 1904.
23. *Auckland Star*, 12 November 1906.
24. *Auckland Star*, 22 December 1906.
25. *Auckland Star*, 11 May 1907.
26. *Otago Witness*, 19 August 1908.
27. *Auckland Star*, 4 August 1908.
28. *Auckland Star*, 20 November 1911.
29. *Auckland Star*, 20 November 1915.
30. *Auckland Star*, 10 September 1915; *Evening Post*, 11 December 1924.
31. *Observer*, 7 March 1903.

CHAPTER 3

1. Peter Shaw, *New Zealand Architecture: From Polynesian Beginnings to 1990*, Hodder & Stoughton, Auckland, 1991, p. 88.
2. Peter Macky, quoted in Bernard Orsman, 'Historic Coolangatta land on the market', *New Zealand Herald*, 20 June 2011.
3. Jennifer Hayman, quoted in Orsman.
4. Peter Macky (with Paul Waite),

REFERENCES

Coolangatta, A Homage, Livadia Publishers Ltd, Auckland, 2010, pp. 103, 107.
5. Macky, p. 153.
6. Shaw, p. 88.
7. Macky, pp. 155, 163, 166; Simon Farrell-Green, *Metro*, June 2007, p. 55.
8. Macky, pp. 103, 135–44.
9. http://www.historic.org.nz/corporate/registersearch/ProfessionalBio/Professional.aspx?CPName=Bamford%2C+Noel http://www.historic.org.nz/corporate/registersearch/ProfessionalBio/Professional.aspx?CPName=Bamford+_amp_+Pierce
10. Farrell-Green, *Metro*, p. 55.
11. Macky, pp. 15, 20.
12. Farrell-Green, *Metro*, p. 57.
13. *New Zealand Herald*, 18 December 2006.
14. Farrell-Green, *Metro*, p. 59.
15. Hamish Keith, 'Auckland heritage suffers the death of a thousand careless cuts', *NZ Listener*, 18 February 2012.
16. Farrell-Green, *Metro*, p. 54; *New Zealand Herald*, 20 June 2011.
17. Farrell-Green, *Metro*, p. 59.
18. Peter Macky (with Paul Waite), *Coolangatta, A Homage*, Livadia Publishers Ltd, Auckland, 2010.
19. Macky, p. 77.
20. *Evening Post*, 21 September 1928.
21. *Auckland Star*, 1 August 1932.
22. *Auckland Star*, 4 April 1919.
23. *Auckland Star*, 7 November 1931.

CHAPTER 4

1. *Auckland Star*, 12 December 1901, 11 March 1902.
2. R.C.J. Stone. 'Myers, Arthur Mielziner – Biography', from the *Dictionary of New Zealand Biography. Te Ara – the Encyclopedia of New Zealand*, updated 1 September 2010. Http://www.TeAra.govt.nz/en/biographies/3m66/1
3. *Auckland Star*, 11 March 1902.
4. *Auckland Star*, 24 December 1902.
5. *Auckland Star*, 17 & 25 September 1902 & 13 November 1902; *Otago Witness*, 1 July 1903 & *Auckland Star*, 3 January 1903.
6. *Auckland Star*, 6 June 1903
7. *Auckland Star*, 6 December 1902.
8. *Auckland Star*, 18, 20 & 22 December 1902.
9. *Auckland Star*, 3 January 1903.
10. *Auckland Star*, 27 December 1902 & 3 January 1903.
11. *Auckland Star*, 27 December 1902 & 3 January 1903.
12. *Auckland Star*, 1 January 1903; *Observer*, 10 January 1903.
13. *Auckland Star*, 10 January 1903.
14. *Auckland Star*, 6 & 7 January 1903.
15. *Auckland Star*, 4 March 1903.
16. *Auckland Star*, 20 February 1907.
17. *Auckland Star*, 22 July 1911 & 8 June 1918.
18. *Auckland Star*, 15 February 1928 & 19 January 1943.
19. Brian Rudman, *NZ Listener*, 7 February 1987, pp. 17–19.
20. *New Zealand Herald*, 17 August 1961.
21. *Auckland Star*, 24 December 1987.
22. L. R. Shelton. 'Kerridge, Robert James – Biography', from the *Dictionary of New Zealand Biography. Te Ara – the Encyclopedia of New Zealand*, updated 1 September 2010; http://www.TeAra.govt.nz/en/biographies/4k10/1; Brian Rudman, 'Goodbye to all that?', 23. *NZ Listener*, 7 February 1987, pp. 17–19.
23. Barry Gustafson, *His Way: A Biography of Robert Muldoon*, Auckland University Press, 2002, p. 426.
24. Dinah Holman, *Historic Places in New Zealand*, 20 March 1988, pp. 5–7.
25. *New Zealand Herald*, 29 December 1987.
26. *New Zealand Herald*, 8 January 1987.
27. *Auckland Star*, 24 December 1987.
28. *New Zealand Herald*, 29 December 1987.
29. *New Zealand Herald*, 29 December 1987.
30. *Auckland Star*, 16 December 1987.
31. Holman, pp. 5–7.
32. *New Zealand Herald*, 8 January 1987.
33. *Auckland Star*, 16 December 1987.
34. *Auckland Star*, 24 December 1987.
35. Holman, pp. 5–7.
36. Holman, pp. 5–7.

CHAPTER 5

1. John Logan Campbell, *Poenamo: Sketches of the Early Days of New Zealand; Romance and Reality of Antipodean Life in the Infancy of a New Colony*, Williams and Norgate, London, 1881, pp. 98-99.
2. http://www.historic.org.nz/TheRegister/RegisterSearch/RegisterResults.aspx?RID=525
3. http://www.historic.org.nz/TheRegister/RegisterSearch/RegisterResults.aspx?RID=2623; Geoffrey Thornton, *Cast in Concrete: Concrete Construction in New Zealand 1850–1939*, Reed Books, Auckland, 1996, p. 26; John Stacpoole, *Colonial Architecture in New Zealand*, A. H. and A.W.Reed, Wellington, 1976, p. 205.
4. http://www.historic.org.nz/corporate/registersearch/ProfessionalBio/wProfessional.aspx?CPName=Arnold%2C+Charles
5. John Stacpoole, pp. 202–03.
6. *Auckland Star*, 22 June 1912.
7. *Auckland Star*, 22 June 1912.
8. *Auckland Star*, 8 August 1908.
9. *Auckland Star*, 22 June & 12 August 1912.
10. *Auckland Star*, 16 August 1912.
11. *Evening Post*, 17 January 1914.
12. *Auckland Star*, 16 January 1914 & 16 January 1915.
13. *Auckland Star*, 19 December 1914 & 22 January 1915.
14. *Auckland Star*, 7 July 1924.
15. *Auckland Star*, 24 March 1915.
16. *Auckland Star*, 13 March 1915.
17. *Auckland Star*, 25 March 1915.
18. *Auckland Star*, 30 November 1915.
19. *Auckland Star*, 4 & 23 November 1918.
20. *Auckland Star*, 22 & 29 November 1918.
21. *Auckland Star*, 10 December 1918 & 6 February 1920.
22. *New Zealand Herald*, 12 November 1920.
23. *New Zealand Herald*, 23 June 1921.
24. *New Zealand Herald*, 7 July 1924.
25. *Auckland Star*, 26 August 1921.
26. *New Zealand Herald*, 11 June 1924.
27. *Auckland Star*, 6 May 1925.
28. Ivy Hutchinson, *New Zealand Herald*, 27 June 1964.
29. *Auckland Star*, 15 December 1890.
30. *Auckland Star*, 24 March 1892, 15 November 1917 & 21 January 1918.
31. *Auckland Star*, 15 March 1961.

CHAPTER 6

1. George Everard Bentley, *The Story of the Old Windmill: gross negligence or rank favouritism on the part of the city council: one of Auckland's oldest landmarks in danger of disappearing*, A. Spencer (printer), Auckland, 1898, p. 3.
2. *New Zealand Herald*, 26 August 1978.
3. John Stacpoole, *Colonial Architecture in New Zealand*, A.H. & A.W. Reed, Wellington, 1976, p. 69.
4. *Daily Southern Cross*, 2 December 1851; *Partington's Windmill*, Lodestar Press, Auckland 1975, (unpaginated).
5. *Daily Southern Cross*, 12 February 1856, *Daily Southern Cross*, 3 October 1866, *New Zealand Herald*, 26 March 1864.
6. *New Zealand Herald*, 26 March 1864.

7. *New Zealand Herald*, 26 March 1864; Bentley, p. 3.
8. *Daily Southern Cross*, 3 October 1866.
9. *Partington's Windmill*.
10. *New Zealand Herald*, 9 January 1873.
11. *New Zealand Herald*, 20 March 1873.
12. *Auckland Star*, 5 March 1875.
13. *Auckland Star*, 27 November 1876.
14. *Partington's Windmill*.
15. *Auckland Star*, 11 August 1887.
16. *Observer*, 4 March 1882.
17. *Auckland Star*, 11 August 1887.
18. *Partington's Windmill*.
19. Bentley, p. 9.
20. Bentley, p. 23.
21. *New Zealand Herald*, 10 September 1898.
22. *Partington's Windmill*.
23. *Auckland Star*, 29 September 1909.
24. *Partington's Windmill*.
25. 'Ruthyn', 'The Old Windmilll', in *The Mirror*, 1 December 1928, p. 49.
26. Ruth Park, *A Fence Around the Cuckoo*, Penguin, Victoria, Australia, 1992, pp. 236–39.
27. 'Old Windmill Preservation Society', brochure, undated, in Auckland Scrapbook, 1949–1952, Auckland Central Library, p. 107.
28. *New Zealand Herald*, 2 November 1945.
29. *New Zealand Parliamentary Debates*, vol. 297, p. 508.
30. *New Zealand Parliamentary Debates*, vol. 299, p. 736.
31. *New Zealand Herald*, 26 August 1978.
32. *New Zealand Herald*, 3 December 1982.
33. *Auckland Star*, 12 September 1938.

CHAPTER 7

1. John Stacpoole. 'Mason, William – Biography', from the *Dictionary of New Zealand Biography. Te Ara – the Encyclopedia of New Zealand*, updated 1 September 2010; http://www.TeAra.govt.nz/en/biographies/1m22/1
2. Stacpoole, *William Mason: The First New Zealand Architect*, Auckland University Press, 1971, pp. 16–17, 38, 47.
3. http://www.aucklanglican.org.nz/?sid=15
4. Stacpoole, 1971, pp. 40–41.
5. Stacpoole, 1971, p. 63; *Auckland Star*, 23 February 1885.
6. *Auckland Star*, 15 December 1884 & 23 February 1885.
7. *Auckland Star*, 28 July 1916.
8. *Auckland Star*, 23 February 1885.
9. *Daily Southern Cross*, 17 February 1874.
10. *Auckland Star*, 9 March 1877.
11. *Taranaki Herald*, 17 April 1883.
12. Charlotte Mary Yonge, *Life of John Coleridge Patterson: Missionary Bishop of the Melanesian Islands*, Macmillan, London, 1878, pp. 96–97.
13. *Auckland Star*, 7 August 1883.
14. *Auckland Star*, 7 August 1883.
15. *New Zealand Herald*, 30 April 1883.
16. *Auckland Star*, 18 April 1883 & 7 August 1883.
17. *Auckland Star*, 7 August 1883.
18. *Auckland Star*, 30 April 1883.
19. *Auckland Star*, 7 August 1883.
20. *Auckland Star*, 22 June 1883.
21. *Auckland Star*, 7 August 1883.
22. *Auckland Star*, 15 December 1884
23. *Auckland Star*, 23 February 1885.
24. *Auckland Star*, 19 February 1885.
25. *Auckland Star*, 23 February 1885.
26. J.A. Froude, *Oceana, or England and her Colonies*, Longmans, Green and Co., New York, 1886, p. 243.
27. *Auckland Star*, 26 March 1885.
28. *Auckland Star*, 28 July 1916; http://www.stpauls.org.nz/welcoming/our_story/wherewevecomefrom1.aspx
29. http://www.historic.org.nz/TheRegister/RegisterSearch/RegisterResults.aspx?RID=650
30. Stacpoole, 1971, p. 86.

CHAPTER 8

1. *New Zealand Herald*, 24 February 1883; *Auckland Star*, 27 October 1882.
2. *New Zealand Tablet*, 18 May 1883.
3. *Poverty Bay Herald*, 12 March 1883.
4. *New Zealand Herald*, 24 February 1883.
5. *New Zealand Herald*, 6 October 1883.
6. *New Zealand Herald*, 6 October 1883.
7. *Auckland Star*, 12 October 1883.
8. *Auckland Star*, 6 October 1883 & 3 November 1883.
9. *Auckland Star*, 23 February 1884; *Otago Daily Times*, 21 February 1884.
10. *New Zealand Herald*, 9 August 1884.
11. *Auckland Star*, 6 May 1884; *New Zealand Herald*, 12 August 1884.
12. *Auckland Star*, 22 June 1885.
13. *Auckland Star*, 16 January 1885.
14. *Auckland Star*, 22 December 1885 & *The Press*, 22 December 1885.
15. *Auckland Star*, 28 January 1886.
16. William Garden Cowie, D.D., Bishop of Auckland, *Our Last Year in Auckland*, 1887, Kegan, Paul, Trench & Co., London, 1888, pp. 293–94.
17. *Auckland Star*, 19 April 1886.
18. *Auckland Star*, 23 January 1886.
19. *Auckland Star*, 10 February 1886.
20. *Auckland Star*, 9 March 1886, 12 April & 29 April 1886.
21. *Auckland Star*, 21 May 1886.
22. *Auckland Star*, 21 December 1887.
23. *New Zealand Herald*, 4 June 1975.
24. *Auckland Star*, 16 September 1887.
25. Ross Fraser, 'Victoria Arcade: Some Auckland Painters of the Fin de Siecle', *Art New Zealand* 20, 1981.
26. *Auckland Star*, 13 October 1886.
27. *Auckland Star*, 30 September & 23 November 1887, and 27 August, 22 November, 16 December 1889.
28. *Auckland Star*, 11 June 1888.
29. *Auckland Star*, 3 December 1888.
30. *Wanganui Herald*, 23 April 1898.
31. *Auckland Directory*, 1931.
32. *The Cyclopedia of New Zealand Auckland Provincial District*, Cyclopedia Company Limited, Christchurch, 1902, p. 49.
33. *Auckland Directory*, 1912.
34. R. Tizard, *The Auckland Society of Arts, 1870–1970: A Centennial History*, printed by Business Printing Works, Auckland, 1972.
35. *Auckland Directory*, 1940–41.
36. John Fields and John Stacpoole, *Victorian Auckland*, John McIndoe, Dunedin, 1973, no. 43 (unpaginated).
37. *New Zealand Herald*, 4 June 1975.
38. John Stacpoole, *Colonial Architecture in New Zealand*, A.H. & A.W. Reed, Wellington, 1976, p. 207.
39. *New Zealand Herald*, 17 October 1977.
40. *New Zealand Herald*, 2 October 1978.
41. *New Zealand Herald*, 17 October 1977.

CHAPTER 9

1. *Poverty Bay Herald*, 30 December 1907.
2. James Mackintosh Bell, *The Wilds of Maoriland*, Macmillan & Co. Ltd, London, 1914, pp. 125–26.
3. Bell, p. 126.
4. Bell, p. 127.
5. *Poverty Bay Herald*, 30 December 1907.
6. Peter Webster, *Rua and the Maori Millennium*, Price Milburn for Victoria University Press, Wellington, 1979, p. 204.
7. Judith Binney, Gillian Chaplin and Craig Wallace, *Mihaia: The Prophet Rua Kenana and His Community at Maungapohatu*, Auckland University Press, Auckland, 1979, pp. 45, 47–49.
8. Bell, p. 128.
9. *Poverty Bay Herald*, 30 December 1907.
10. *Poverty Bay Herald*, 27 March 1908.
11. Binney et al., 1979, pp. 47–49.
12. *Marlborough Express* (reprinted from *New Zealand Herald*), 13 April 1916.
13. *Poverty Bay Herald*, 31 December 1908.
14. Webster, 1979, p. 236.
15. *Marlborough Express* (reprinted from *New Zealand Herald*), 13 April 1916.

REFERENCES

16. Judith Binney. 'Rua Kenana Hepetipa – Biography', from the *Dictionary of New Zealand Biography. Te Ara – the Encyclopedia of New Zealand*, updated 4 July 2012. http://www.TeAra.govt.nz/en/biographies/3r32/1
17. Binney et al. pp. 76, 78, 144.
18. Binney et al., pp. 136, 143–44.
19. Binney et al., p. 185.

CHAPTER 10

1. *Taranaki Herald*, 25 August & 1 September 1852; Clutha Leader, 15 February 1895.
2. *Taranaki Daily News*, 9 July 1938; A.B. Scanlan, *Historic New Plymouth*, A.H. & A.W. Reed, Wellington, 1968, p. 38.
3. *Taranaki Herald*, 28 March 1868 & 7 October 1903.
4. *Taranaki Herald*, 9 September 1901; A.B. Scanlan, p. 38; *Taranaki Herald*, 7 October 1903.
5. Murray Moorhead, *Colonial Tales of Old New Plymouth*, Zenith Publications, New Plymouth, 2005, pp. 78, 81.
6. *Taranaki Herald*, 7 October 1903.
7. Jeremy Salmond, *Old New Zealand Houses 1800–1940*, Heinemann Reed, 1986, pp. 51, 115.
8. Salmond, pp. 27–29.
9. *Taranaki Herald*, 1 September 1852 & 12 October 1883.
10. *Taranaki Herald*, 10 August 1870.
11. *Taranaki Herald*, 17 November 1893, 9 September 1901 & 7 October 1903.
12. *Taranaki Herald*, 12 February, 8 May 1890 & 7 May 1903.
13. *Taranaki Herald*, 24 April 1902 & 29 December 1902.
14. *Taranaki Herald*, 21 & 28 August 1903, 26 August 1928.
15. Moorhead, pp. 78, 81; *Taranaki Herald*, 7 October & 3 December 1903.
16. *Taranaki Herald*, 7 October 1903.
17. *Taranaki Daily News*, 9 July 1938; Moorhead, pp. 78, 81.

CHAPTER 11

1. http://www.waiapu.com/about-us/history/
2. *Daily Telegraph*, 20 December 1888.
3. Ian J. Lochhead. 'Mountfort, Benjamin Woolfield – Biography', from the *Dictionary of New Zealand Biography. Te Ara – the Encyclopedia of New Zealand*, updated 1 September 2010 URL: http://www.TeAra.govt.nz/en/biographies/1m57/1
4. Ian J. Lochhead, *A Dream of Spires: Benjamin Mountfort and the Gothic Revival*, Canterbury University Press, Christchurch, 1999, pp. 162, 164.
5. *Daily Telegraph*, 20 December 1888.
6. *Daily Telegraph*, 20 December 1888; Lochhead, 1999, p. 164.
7. *Daily Telegraph*, 20 December 1888.
8. *Daily Telegraph*, 20 December 1888
9. *Daily Telegraph*, 20 December 1888.
10. Lochhead, 1999, pp. 164, 165, 168.
11. Lochhead, 1999, p. 163.
12. *Poverty Bay Herald*, 10 August 1904.
13. *Poverty Bay Herald*, 9 August 1904; *Manawatu Standard*, 9 August 1904.
14. Lochhead, 1999, p. 172.
15. Lochhead, 1999, p. 172.
16. *Waiapu Church Gazette*, 1 May 1931; Lochhead, 1999, p. 163.
17. http://www.hawkesbaytoday.co.nz/news/st-pauls-celebrates-150-yearsin-style/991777/; Marty Sharpe, 'What Napier can teach Christchurch about earthquake recovery', 26 March 2011, in http://stuff.co.nz/dominion-post/
18. Michael Dunn, 'Roland Hipkins & Renaissance: A Little-Known Painter Reconsidered', *Art New Zealand* 112, Spring 2004, pp. 88–91, 99.
19. http:www.napiercathedral.org.nz/history.php

CHAPTER 12

1. Patricia Burns, *Te Rauparaha: A New Perspective*, A.H. & A.W. Reed, Wellington, 1980, p. 288; Eric Ramsden, *Rangiatea: The Story of the Otaki Church, its First Pastor and its People*, A.H. & A.W. Reed, Wellington, 1951, p. 11.
2. Ramsden, pp. 57, 61; www.http://rangiatea.natlib.govt.nz/OctaviusE.htm
3. Ramsden, pp. 60–61.
4. *New Zealander*, 4 July 1846.
5. Ramsden, pp. 37, 39, 62, 109–110, 153; Burns, p. 288.
6. Burns, p. 288.
7. Ramsden, pp. 110, 145.
8. Richard A. Sundt, *Whare Karakia: Maori Church Building, Decoration and Ritual in Aotearoa New Zealand, 1834–1863*, Auckland University Press, Auckland, 2010, pp. 112–113.
9. Ramsden, pp. 127, 145–47.
10. Sundt, p. 112.
11. Sundt, p. 113.
12. *Wellington Independent*, 26 December 1849.
13. Ramsden, pp. 122, 143; www.http://rangiatea.natlib.govt.nz/OctaviusE.htm
14. Sundt, p. 113.
15. www.http://rangiatea.natlib.govt.nz/BuildingE.htm
16. Ramsden, p. 62.
17. Burns, p. 289.
18. Sundt, pp. 113, 121.
19. Ramsden, p. 309.
20. Sundt, pp. 116, 121; Sarah Treadwell, *Rangiatea Revisited*, School of Art History, Classics and Religious Studies, Victoria University of Wellington, 2008, p. 33; 'Maori Church at Otaki', Progress, 1 December 1909, p. 64.
21. Ramsden, pp. 316, 319–320.
22. Ramsden, pp. 11, 178, 322.
23. Ramsden, p. 338.
24. www.http://rangiatea.natlib.govt.nz/FireE.htm; Treadwell, p. 26.
25. Treadwell, p. 43.

CHAPTER 13

1. *Auckland Star*, 11 December 1907.
2. *Auckland Star*, 11 December 1907.
3. *The Press*, 3 March 1873.
4. 'A Brilliant Scene', *Evening Post*, 11 December 1907.
5. 'Parliament's Home Destroyed', *Evening Post*, 11 December 1907.
6. http://www.historic.org.nz/en/TheRegister/RegisterSearch/RegisterResults.aspx?RID=217.
7. Anna Crighton. 'Clayton, William Henry – Biography', from the *Dictionary of New Zealand Biography. Te Ara – the Encyclopedia of New Zealand*, updated 1 September 2010; URL: http://www.TeAra.govt.nz/en/biographies/2c20/1.
8. *Auckland Star*, 11 December 1907.
9. *Auckland Star*, 11 December 1907.
10. 'Parliament's Home Destroyed', *Evening Post*, 11 December 1907.
11. *Auckland Star*, 11 December 1907.
12. 'The Library', *Evening Post*, 11 December 1907.
13. 'A Brilliant Scene', *Evening Post*, 11 December 1907.
14. 'No Insurance on the Building', *Evening Post*, 11 December 1907; *Auckland Star*, 11 December 1907.
15. 'In the Past', *Evening Post*, 11 December 1907.
16. *Auckland Star*, 11 December 1907.
17. http://www.parliament.nz/mi-NZ/AboutParl/HowPWorks/FactSheets/3/c/c/00PlibJMPBG1-History-of-Parliament-s-buildings-and-grounds.htm.
18. *New Zealand Parliamentary Debates*, vol. 143, 1908, p. 1.
19. *New Zealand Parliamentary Debates*, vol. 143, 1908, p. 190.
20. *New Zealand Parliamentary Debates*, vol. 143, 1908, pp. 302, 330, 335.

CHAPTER 14

1. http://www.historic.org.nz/TheRegister/RegisterSearch/RegisterResults.aspx?RID=7201
2. *The Colonist*, 5 March 1858.
3. *The Colonist*, 3 August 1858.
4. Edwin Hodder, *Memories of New Zealand Life*, Longman, Green, Longman & Roberts, London, 1862, p. 36.
5. *Nelson Examiner and New Zealand Chronicle*, 27 August 1859.
6. *The Colonist*, 16 November 1858.
7. John Stacpoole, *Colonial Architecture in New Zealand*, A.H. & A.W. Reed, Wellington, 1976, p. 69; Anne Marchant, 'Bury, Maxwell – Biography', from the *Dictionary of New Zealand Biography. Te Ara – the Encyclopedia of New Zealand*, updated 1 September 2010. http://www.TeAra.govt.nz/en/biographies/2b52/1
8. Marchant. http://www.TeAra.govt.nz/en/biographies/2b52/1
9. Stacpoole, p. 71.
10. *The Colonist*, 30 August 1859.
11. *The Colonist*, 30 August 1859.
12. *The Colonist*, 30 August 1859.
13. *Nelson Examiner and New Zealand Chronicle*, 27 August 1859.
14. http://www.historic.org.nz/TheRegister/RegisterSearch/RegisterResults.aspx?RID=7201NZHPT; http://www.nelsoncitycouncil.co.nz/albionsquare/?calendardate=2012-10-1
15. Hodder, pp. 35–36.
16. Hodder, p. 36.
17. *The Colonist*, 16 August 1861.
18. http://www.historic.org.nz/TheRegister/RegisterSearch/RegisterResults.aspx?RID=7201
19. *The Colonist*, 26 February 1906.
20. *The Colonist*, 3 August 1858; information kindly provided by Ted Anderson.
21. http://www.nelsoncitycouncil.co.nz/albionsquare/?calendardate=2012-10-1
22. http://www.historic.org.nz/TheRegister/RegisterSearch/RegisterResults.aspx?RID=7201NZHPT; http://www.nelsoncitycouncil.co.nz/albionsquare/?calendardate=2012-10-1
23. John Wilson, *Thematic Historical Overview of Nelson City*, October 2011, p. 31, http://www.nelsoncitycouncil.co.nz/assets/About-nelson/downloads/1176312-thematic-historical-overview-of-nelson-city-OCT2011.pdf

CHAPTER 15

1. http://christchurchcitylibraries.com/Heritage/Publications/ChristchurchCityCouncil/ArchitecturalHeritage/LegacyofThomasEdmonds/LegacyofThomasEdmonds.pdf, p. 2.
2. *Through the Changing Years*, T.J. Edmonds Ltd, 1929 [unpaginated].
3. *Through the Changing Years.*
4. http://christchurchcitylibraries.com/Heritage/Publications/ChristchurchCityCouncil/ArchitecturalHeritage/LegacyofThomasEdmonds/LegacyofThomasEdmonds.pdf, p. 5.
5. G.A. Tait (editor), *Manufacturing in New Zealand*, Cranwell Publishing Co Ltd, Auckland, 1959, B-16.
6. http://www..historic.org.nz/TheRegister/RegisterSearch/RegisterResults.aspx?RID=7486 .
7. http://christchurchcitylibraries.com/Heritage/Publications/ChristchurchCityCouncil/ArchitecturalHeritage/LegacyofThomasEdmonds/LegacyofThomasEdmonds.pdf, pp. 4, 7.
8. Jenny Chamberlain, 'Old Faithfuls: The Best of the Best Sellers', *North & South*, April 1990, p. 93.
9. http://www.nzhistory.net.nz/media/photo/edmonds-cookbook
10. http://christchurchcitylibraries.com/Heritage/Publications/ChristchurchCityCouncil/ArchitecturalHeritage/LegacyofThomasEdmonds/LegacyofThomasEdmonds.pdf, p. 11.
11. ait, B-16.
12. *Your Visit to Edmonds*, T.J. Edmonds Ltd, Christchurch, 1957 unpaginated
13. http://christchurchcitylibraries.com/Heritage/Publications/ChristchurchCityCouncil/ArchitecturalHeritage/LegacyofThomasEdmonds/LegacyofThomasEdmonds.pdf, p. 3.
14. Tait, B-16; *Your Visit to Edmonds.*
15. Tait, B-15.
16. Chamberlain, p. 92.
17. Bruce Ansley, *NZ Listener*, 12 November 1990, p. 20.
18. Ansley, p. 20.

CHAPTER 16

1. *Nelson Examiner and New Zealand Chronicle*, 13 November 1847.
2. *New Zealand Spectator and Cook's Strait Guardian*, 6 August 1864.
3. *New Zealand Herald*, 6 August 1864.
4. *New Zealand Herald*, 9 June 1865; *Daily Southern Cross*, 10 June 1865.
5. *Otago Daily Times*, 5 August 1867.
6. *Lyttelton Times*, 11 January 1866.
7. *Christchurch Star*, 4 August 1876.
8. *The Press*, 5 May 1877.
9. http://www.historicplaces.org.nz/placesvisit/canterbury/lyttelttontimeballstation.aspx
10. *Evening Post*, 28 April 1877.
11. *New Zealand Herald*, 8 August 1883.
12. http://www.historicplaces.org.nz/en/placesToVisit/canterbury/LytteltonTimeballStation/History.aspx.
13. *The Press*, 28 May, 1890.
14. *The Press*, 16 July 1891.
15. *Evening Post*, 17 May 1933.
16. http:www.bobmckerrow.blogspot.co.nz/2011/06/damage-to-timeballstation-in-lyttelton.html http://www.historic.org.nz/TheRegister/RegisterSearch/RegisterResults.aspx?RID=1840
17. *Wanganui Herald*, 9 March 1908.
18. *Auckland Star*, 16 May 1910.
19. http://www.historicplaces.org.nz/en/placesToVisit/canterbury/LytteltonTimeballStation/History.aspx.
20. Amanda Cropp, 'Calling Time', *Heritage New Zealand*, Winter 2011.
21. http://www.historicplaces.org.nz/placesvisit/canterbury/lytteltontimeballstation.aspx
22. http://www.historicplaces.org.nz/placesvisit/canterbury/lytteltontimeballstation.aspx & http://tvnz.co.nz/national-news/1mdonation-saves-lyttelton-timeballstation-5447766

CHAPTER 17

1. *Otago Daily Times*, 15 July 1879.
2. *Otago Daily Times*, 14 March 1879, 15 July 1879.
3. *Otago Daily Times*, 15 October 1898.
4. *Otago Daily Times*, 14 March 1879, 15 July 1879.
5. *The Press*, 15 July & *Otago Daily Times*, 7 November 1879.
6. *Otago Daily Times*, 15 July 1879.
7. *Otago Daily Times*, 15 July 1879.
8. *Otago Daily Times*, 15 July 1879.
9. John Stacpoole, *Colonial Architecture in New Zealand*, A.H. & A.W. Reed, Wellington, 1976, p. 134.
10. *Otago Daily Times*, 13 August 1880.
11. *Otago Daily Times*, 23 July 1881.
12. *Otago Daily Times*, 13 July 1882.
13. *Otago Daily Times*, 15 June 1883.
14. *Otago Daily Times*, 12 August 1885.

15. *Otago Daily Times*, 27 May 1886.
16. *Otago Daily Times*, 22 February 1888.
17. *Otago Daily Times*, 22 March 1888.
18. Jonathan Mane-Wheoki, 'Lawson, Robert Arthur – Biography', from the *Dictionary of New Zealand Biography. Te Ara – the Encyclopedia of New Zealand*, updated 1 September 2010. http://www.TeAra.govt.nz/en/biographies/2l5/1
19. Barbara Brookes. 'King, Frederic Truby – Biography', from the *Dictionary of New Zealand Biography. Te Ara – the Encyclopedia of New Zealand*, updated 1 September 2010. http://www.TeAra.govt.nz/en/biographies/2k8/1.
20. *Cyclopedia of New Zealand*, 1906, pp. 147–48.
21. E.C. Grayland, *New Zealand Disasters*, A.H. & A.W. Reed, Wellington, 1957, pp. 155–57.
22. Patrick Evans. 'Frame, Janet Paterson - Hospitalisation and publication', from the *Dictionary of New Zealand Biography. Te Ara - the Encyclopedia of New Zealand*, updated 6-Dec-10 URL: http://www.TeAra.govt.nz/en/biographies/6f1/2.
23. Heather Bauchop, 'Seacliff Lunatic Asylum Site', 27 January 2012, http//www.historic.org.nz/~/media/Corporate/Files/Register/.../Seacliff.ashx, pp. 3, 4, 18–20.

CHAPTER 18

1. Ewan Johnston, 'A Valuable and Tolerably Extensive Collection of Native and other Products: New Zealand at the Crystal Palace', in Jeffrey A. Auerbach and Peter H. Hoffenberg (editors), *Britain, the Empire, and the World at the Great Exhibition of 1851*, Ashgate Publishing Company, c.2008, Aldershot, England & Burlington, Vermont, pp. 78-84.
2. *Otago Witness*, 19 December 1862; *Otago Daily Times*, 18 February 1864, *Lyttelton Times*, 17 January 1865.
3. *Otago Daily Times*, 25 April 1864.
4. John Stacpoole, *William Mason: The First New Zealand Architect*, Auckland University Press, Auckland, 1971, pp. 82–83.
5. *Otago Witness*, 25 November 1864.
6. *New Zealand Spectator and Cook's Strait Guardian*, 24 February 1864.
7. *Otago Daily Times*, 18 February 1864; *New Zealand Spectator and Cook's Strait Guardian*, 24 February 1864.
8. *New Zealand Spectator and Cook's Strait Guardian*, 24 February 1864.
9. *New Zealand Spectator and Cook's Strait Guardian*, 24 February 1864.
10. *Otago Witness*, 25 November 1864.
11. Stacpoole, p. 82; *Otago Witness*, 25 November 1864.
12. *New Zealand Spectator and Cook's Strait Guardian*, 24 February 1864.
13. *Lyttelton Times*, 17 January 1865.
14. *North Otago Times*, 19 January 1865.
15. James Hamilton, *London Lights: The Minds that Moved the City that Shook the World, 1805–51*, John Murray, London, 2007, p. 305.
16. *New Zealand Spectator and Cook's Strait Guardian*, 24 February 1864.
17. Stacpoole, 1971, p. 83.
18. *Otago Daily Times*, 8 May 1865.
19. *Otago Daily Times*, 6 May 1865.
20. John H. Angus, *A History of the Otago hospital board and its Predecessors*, Otago hospital board, Dunedin, 1984, pp. 26–27, 29, 33.
21. Stacpoole, p. 84.
22. Angus, pp. 33–34.
23. Angus, pp. 35, 83.
24. *Dunedin Contextual Thematic History*: 6.1 Political, at http://dcc.squiz.net.nz/_data/assets/pdf_file/
25. Angus, p. 197.

CHAPTER 19

1. http://www.teara.govt.nz/en/1966/invercargill/1.
2. *Southland Times*, 8 August 1893.
3. Paul Sorrell, *Muruhiku: The Southland Story*, Southland to 2006 Book Project Committee, Invercargill, 2006, pp. 148–49.
4. J.O. P. Watt, *Centenary of Invercargill Municipality 1871–1971*, printed by Times Printing Service, Invercargill, 1971, p. 156.
5. *Southland Times*, 8 August 1893.
6. *Southland Times*, 8 August 1893.
7. *Otago Witness*, 22 May 1901.
8. David McGill, *Landmarks: Notable Historic Buildings of New Zealand*, Godwit, Auckland, 1997, p. 246.
9. *Southland Times*, 8 August 1893.
10. Watt, p. 156.
11. Watt, p. 157.
12. Sorrell, p. 316.
13. Watt, p. 157; *Town Belt Management Plan, Invercargill*, [not dated], 5.23.
14. Watt, p. 158.
15. Watt, p. 158.
16. http://www.waymarking.com/waymarks/WM720G_Town_Clock_Invercargill_New_Zealand.
17. http://www.historic.org.nz/TheRegister/RegisterSearch/RegisterResults.aspx?RID=1087.
18. http://www.historic.org.nz/TheRegister/RegisterSearch/RegisterResults.aspx?RID=4510
19. McGill, p. 236; http://www.historic.org.nz/TheRegister/RegisterSearch/RegisterResults.aspx?RID=394

CHAPTER 20

1. http://www.historic.org.nz/TheRegister/RegisterSearch/RegisterResults.aspx?RID=7777.
2. *History of the Southland Hospitals and Board, 1861–1968*, Invercargill, 1968, p. 1.
3. *History of the Southland Hospitals and Board, 1861–1968*, Invercargill, 1968, p. 5.
4. *History of the Southland Hospitals and Board, 1861–1968*, Invercargill, 1968, pp. 6, 7.
5. Jill Corson, *Urban Design Strategy: South Invercargill*, June 2010, p. 3.3.2.
6. Frances Porter, in *Historic Buildings of New Zealand: North Island* (editor, Frances Porter), Cassell Ltd., Auckland, 1979, pp. 179–181.
7. http://www.historic.org.nz/TheRegister/RegisterSearch/RegisterResults.aspx?RID=7777.
8. *Southland Times*, 25 November 1876.
9. http://www.historic.org.nz/TheRegister/RegisterSearch/RegisterResults.aspx?RID=7777.
10. Jonathan Mane-Wheoki, 'Burwell, Frederick William (1846–1915)' in Jane Thomson (editor), *Southern People: A Dictionary of Otago Southland Biography*, Dunedin, 1998, p. 74.
11. *Southland Times*, 9 March 1897.
12. *Southland Times*, 28 November, 1900; 6. *History of the Southland Hospitals and Board, 1861–1968*, Invercargill, 1968, pp. 37–42.
13. John McKenzie and Edmund Richardson Fitz Wilson.
14. http://www.historic.org.nz/TheRegister/RegisterSearch/RegisterResults.aspx?RID=7777.
15. *Southland Times*, 28 November 1900.
16. *Southland Times*, 20 September 1904.
17. *History of the Southland Hospitals and Board, 1861–1968*, Invercargill, 1968, pp. 37–42.
18. http://www.historic.org.nz/TheRegister/RegisterSearch/RegisterResults.aspx?RID=7777

INDEX

1840–1940 New Zealand Centennial 8
1865 New Zealand Exhibition Building 8, 158–165
1870 Paris Exhibition 56
1913–14 Auckland Exhibition 38

A

'A Runaway Girl' 45, 47
Abbot, R. Atkinson 30
Abbott, Robert Henry 42, 48
Acacia Cottage, Auckland 8, 54
Admiralty House, Auckland 8, 15, 28–33
Admiralty House Bill, 1898 30
Allen, Jim 86
Ambler, F.N. 68
Anthroposophical Society 86
Anzac Avenue, Auckland 55, 58
Aotea Centre, Auckland 51
Archey, Gilbert 68
Arnold, Charles Le Neve 30, 33
Arts and Crafts 84
Aston Hall, Birmingham 130
Athenaeum, Invercargill 171
Atkinson, Robert 83
Auckland
  Admiralty House 8, 15, 28–33
  Coolangatta 14–15, 34–39
  His Majesty's Theatre and Arcade 10, 12, 36, 40–51, 62
  Kilbryde 8, 12, 15, 52–59.
  Partington's Mill 8, 12, 36, 60–69
  St Paul's 62, 70-77
  Victoria Arcade 10, 44, 58, 62, 78–87
Auckland Anglican Diocese 83
Auckland Art Gallery 12, 39, 82, 85
Auckland Benevolent Society 83
Auckland Board of Education 83
Auckland Central Post Office 10, 12, 173
Auckland City Council 51
Auckland City Council District Plan 38
Auckland Civic Trust 49, 69
Auckland College 83
Auckland Customhouse 12, 85
Auckland Education Board 58
Auckland Grammar School 30, 83
Auckland Harbour Board 29, 57, 147
Auckland Hospital Board 58

Auckland Institute of Architects 82
Auckland Library 12, 82, 85
Auckland Political Financial Reform Association 83
Auckland Society of Arts 10, 84
Auckland Supreme Court 12
Auckland Teachers' Training College 58
Auckland Town Hall 42
Auckland War Memorial Museum 39
Augustus Terrace, Auckland 75
Australian Cream of Tartar Organisation 139
Avondale Hospital, Auckland 157

B

Balmoral, Scotland 154
Bamford, F. (Frederick) Noel 36, 39
band rotunda, Invercargill 171–172
Bank of New Zealand, Auckland 82, 86
Barr, John 68
Bartley, Alva 85
Bartley, Edward 38
Beagle, HMS 146
Beatson, William 128
Beehive, Wellington 125
Belich, James 21
Bell, Dr James 91
Bell Block, Taranaki 99
Bella Vista boarding house, Auckland 33
Bellamy's, Wellington 123
Bennett, Right Reverend F.A. 117
Bentley, George Everard 67
Berry, Alexander 36
Birthright 86
Bishopscourt, Auckland 38
Bledisloe, Lord 8
Blomfield, Charles 83–84
Bluebird Foods 140
Bluff 168, 173
Bluff Hill, Napier 107
Brackenhirst, Taranaki 99
Brierley Cromwell Properties 140–141
British Antarctic Expedition 148
Britomart Transport Centre, Auckand 10, 12, 173
Brooklands, New Plymouth 181
Brown, William 54
Burton Bros. 84

Burwell, Frederick W. 171, 178
Bury, Maxwell 128–130, 133
Bycroft, John 62

C

Caledonia Hotel, Auckland 69
Campbell and Ehrenfried 42
Campbell, Sir John Logan 8, 54–59, 122, 125
Campbell's Point, Auckland 56–59
Cane, Thomas 146
Canterbury College 106
Canterbury Earthquake Recovery Act (Cera) 13
Canterbury Earthquake Royal Commission 14
Canterbury earthquakes, 2010 and 2011 141, 144.
Canterbury Library 13
Canterbury Museum 106
Canterbury Provincial Council 146
Canterbury Provincial Council Buildings 13–14, 106, 125
Carrington Psychiatric Hospital 62
Cashmere, Christchurch 15
Catholic Cathedral of the Blessed Sacrament, Christchurch 13
Chapel of the Little Sisters of the Poor, Auckland 12
Chapman, Bruce 149
Charles, Prince 42, 50
Cherry Farm Hospital, Waikouaiti 157
Chief Post Office, New Plymouth 173
Christ Church, Nelson 129
Christchurch
  T.J. Edmonds Ltd Factory 134–141
Christchurch Boys' High School 139
Christchurch Cathedral 13, 106
Christchurch City Council 141
chronometer 144
Churton, Rev. J.F. 72
City Life Apartments, Auckland 51
Civic Theatre, Auckland 12
Clarendon Hotel, Christchurch 13
Clayton, William 121, 123, 164, 171
Clock Tower, Christchurch 140–141
Colenso, William 112
Colonial Sugar Refinery Office, Auckland 30

INDEX

Columbia Market, London 9
Commercial Bay, Auckland 54
concrete 56
Cook, Captain James 9, 146, 191
Coolangatta, Auckland 14–15, 34–39
Cornwall Park, Auckland 8, 54, 55
Cotton, William 73
Courtenay Street, New Plymouth 101
Coutts, Morton 37
Covent Garden, London 162
Cowan, James 20, 21
Cowie, Bishop William G. 82
Crystal Palace, London, 129, 160, 162–164
Custom House, Wellington 144

D
Darwin, Charles 146
Davies, Windsor 49
Davis, Sir Ernest 59
de Clere, F. 116
Dee Street Hospital, Invercargill 174–181
Despard, Colonel 7, 20–25
Devon Line, New Plymouth 98
Devonport Post Office 173
Diamond Jubilee 176
'Dick Whittington' 47
Dominion Breweries 37
Duff, Alison 86
Dunedin
    1865 New Zealand Exhibition Building 8, 158–165
    Seacliff Lunatic Asylum 150–157
Dunedin City Council 173
Dunedin Hospital 164
Dunn, Michael 109

E
Edmonds, Thomas John 136–141
Edmonds' Cookery Book 137, 139
Edward Dent & Co., London 147
Edward Mahoney & Son 42, 56
Edward VII, King 42, 90
Egmont House, New Plymouth 98–99
Eiffel Tower, Paris 148
Elam School of Art 67, 84
Emily Place, Auckland 58
Euston Arch, London 9

F
façadism 12, 15
Fairburn, Rex 86
Fanshawe, Admiral 31
Feldon, William H. 172
Ferry Building, Auckland 147
Ferry Road, Christchurch 136
Ferrymead Heritage Park 140
Finch, Walter P. 109
fires, Wellington 125
FitzRoy, Robert 18, 146

Fort Britomart, Auckland 72
Foster, Jessie 36
Fowler, Charles 162
Frame, Janet 157
Frankton 125
Freyberg, Governor General Bernard 117
Friends of the Earth 86
Froude, James Anthony 8, 77

G
Gala Street, Invercargill 173
Gay Publishing Collective 86
General Assembly Library 122
General Government Building, Invercargill 171
Geological Survey 155
George Court's Building, Auckland 84
Gisborne 90
Gladstone Road, Auckland 56
Glenalvon boarding house, Auckland 33
Gloucester, Duke of 181
gold, discovery of 176
Goldie, Charles F. 69, 83–84
Goldie, David 69
Goldsbro' and Wade 85
Goodman Fielder 141
Government Buildings, Wellington 122
Government House, Auckland 100
GPS (Global Positioning System) 148
Grafton Bridge, Auckland 42
Great Exhibition of the Works of Industry of all Nations 133, 160
Great Northern Brewery, Auckland 30
Greenwich 144
Grey, Sir George 18, 21, 28, 100, 115, 117, 164
Gustafson, Barry 48
Guthrie, John Steele and Maurice James 139
Gwalior 98

H
Hadfield, Bishop Octavius 112
Hangaroa River 92
Harris, Chantrey 98
Harrison, John 144
Hastings Post Office 173
Hawke's Bay earthquake 108–109, 181
Hay, J. A. Louis 109
Hector, Dr James 155
Henderson, Louise 86
Hiona 8, 88–95
Hipkins, Roland 109
Hirst, Thomas and Grace 98
Hirurharama 93, 95
His Majesty's Arcade and Theatre 10, 12, 36, 40–51, 62
Historic Places Act, 1954 9
Historic Places Trust 141
Hobson, William 72

Hochstetter, Ferdinand 130–132
Hodges, William 181
Hodgson, Terence E.R. 15
Holly Lea, Christchurch 130
Hone Heke Pokai 18, 21, 74
hospitals, early 178
House of Representatives 122
Humphreys, Barry 50
Hungerford Market, London 162

I
Iharaira 90
Indian and Colonial Exhibition, London 82
influenza epidemic 58
Invercargill
    Dee Street Hospital 174–181
    Post Office 166–173
Invercargill Post Office 166–173
Invercargill Water Tower 173
Islington, Lord 47
Ivey Hall, Canterbury 130

J
J.C. Williamson 45, 48
Jacobean style 128, 130
Jean Batten Place 83
Jermyn Street, Auckland 33, 55
Jones, Gerald 84
Jones, Owen 163

K
kahu papura 117
Kapiti Coast 112
Kawakawa River 20
Kawiti 7, 18–24
Kerridge Odeon 48
Kerridge, Robert 48
Kew, Invercargill 181
Kilbryde, Auckland 8, 12, 15, 52–59
Kilcaldie's, Wellington 13
King, Frederic Truby 156
King's College Memorial Chapel 30
Kneebone, Francis 42
Knight, C.R. 68
Koroki, King 117
Kororareka 18
Kuhtze, Joseph 37

L
Lambeth Conference 117
Langham Hotel Auckland 69
Langley, Eve 68
Larnach Castle, Dunedin 154
Lawson, Robert 154
Lee-Johnson, Eric 86
Legislative Council 122
Linwood High School 141
Liverpool, Earl of 47
Lochhead, Ian 14, 107

189

Logan Bank 55–56
Longbeach, Ashburton 15
longitude 144
Longwood, Wairarapa 15
Lutyens, Edwin 38
Lyttelton 136
　Timeball Station 142–149
Lyttelton Gas Company 147
Lyttelton Harbour Board 148
Lyttelton Timeball Station 142–149

M

Macalister, Molly 86
Mackenzie, Stuart 50
Macky, Peter 39
Mahlstick Club 83
Mahoney, Thomas 12, 56
Mair, J.T. 172
mangopare 115
Maple Furnishing, Auckland 69
Marsden, Rev. Samuel 104
Martin, Josiah 85
Mason and Wales 165
Mason, William 64, 72, 162, 164
Masonic Hall, Invercargill 178
Masonic Hall, Nelson 129, 133
Massey, William 125
Matthews, Bishop Victoria 13
Maungapohatu 8, 90–95
Mawson, Douglas 148
McDonald, Allan 15
McKenzie and Wilson 180
Mechanics Bay, Auckland 55
Metropolis, Auckland 12
Miller, Myrtle 69
Ministry of Works 133, 148
Morrieson, Ronald Hugh 181
Motukorea (Brown's Island) 56
Mould, Colonel Thomas 74
Mount Victoria, Wellington 147
Mountfort, Benjamin W. 104, 130, 156
Mt. Eden Prison, Auckland 94
Muldoon, Robert 48
Museum of Transport and Technology, Auckland 69
Mutikotiko 114
Myers, Arthur Mielziner 42
'My Fair Lady' 48, 50

N

Napier
　St John's Cathedral 8, 102–109
Napier earthquake 14
Napier Municipal Theatre 51
National Historic Places Trust 9
National Historic Places Trust Bill 8, 69
National Trust, Australia 10
National Trust, United Kingdom 9
Nazarites 90

Neligan, Bishop 38
Nelson
　Provincial Government Building 126–133
Nelson Institute 129
Nelson Provincial Government Building 126–133
Nelson Provincial Museum 133
New Plymouth
　Round House 8, 96–101
New Zealand Army 148
New Zealand Company 98, 128
New Zealand Entertainment Artistes' Benevolent Fund 50
New Zealand Foundation for Peace Studies 86
New Zealand Historic Places Trust 9, 12, 13, 48, 133, 148–149, 173, 181
New Zealand Insurance Company 80
New Zealand Mean Time 147
New Zealand Standard Time 147
Ngata, Sir Apirana 116
Ngati Porou 116
Ngati Raukawa 112, 116
Ngati Toa 112
Nimrod 148
Northland
　Ruapekapeka 7, 18–25

O

Official Bay, Auckland 76
Ohaeawai 18
Ohau River 114
Okukari, Queen Charlotte Sound 112
Old Government House, Auckland 77
Old Windmill Preservation Society 68–69
Olveston, Dunedin 130
O'Brien, George 164
O'Connell Street, Auckland 54
Orakei, Auckland 54
Otago Museum 165
Otago Provincial Council 161
Otaki 112
　Rangiatea 110–117
Owen, Bishop 117

P

Pacer Kerridge 48, 51
Park, Ruth 68
Parliament Buildings, Wellington 8, 92, 118–125
Parliamentary Library 122
Parnell Rose Gardens 59
Parr, Sir James 56
Partington Mill Restoration Society 69
Partington, Charles Frederick 62–66
Partington, Edward 66
Partington, Henry 64
Partington, Joseph 66–68
Partington, Maria 69

Partington's Mill 8, 12, 36, 60–69
Patuone 21
Paxton, Joseph 162
Payton, Edward 83–84
Penn (Pennsylvania) Station, New York 9
Pierce, A.P. Hector 38
Pitt Street Methodist Church 62
Pitt, William 42, 44, 51
Plunket, Lord 125
Point Britomart, Auckland 7, 74–76
Pompallier, Bishop 18
Port Chalmers Observatory 144
Post office, Invercargill 8
Post offices, New Zealand 168
prefabricated houses 100
Pregnancy Help 86
Prickett, Nigel 25
Princess Theatre, Melbourne 44
Provincial Council buildings 9
Provincial Councils 176

Q

Queen Street, Auckland 44
Queen Street, fire 8
Queen's Head Tavern, Auckland 13
Quest Invercargill 173

R

Radiant Hall, Christchurch 140–141
Radiant Health Club, Christchurch 139
Ranfurly, Lord 180
Rangiatea 110–117
Rangitata 92
Ra'iatea 114, 117
Rationalist Association 86
Regent Theatre, Christchurch 13
Regent Theatre, Dunedin 50
Ringatu Church 90
Riverton 178
Robinson, John Perry 130–133
'Rocky Horror Picture Show' 48
Rosicrucian Order 86
Ross, David 165
Round House, New Plymouth 8, 96–101
Rua Kenana Hepetipa 8, 90–95
Ruapekapeka 7, 18–25
Ruatahuna 90
Ruatoria 116
'Run For Your Wife' 49
Ruskin, John 107

S

Salvation Army Citadel (Congress Hall) 12
'San Toy' 47
Savage, H. Clinton 84
Savage, Michael Joseph 59
Scherff, Mrs 33
School of Architecture, The University of Auckland 39

INDEX

Schoon, Theo 86
Scotch Baronial style 154
Scott, Captain Robert Falcon 148
Seabrook Fowlds Motors, Auckland 69
Seacliff Lunatic Asylum 150–157
Seddon, Richard John 29, 31, 120
Selwyn, Bishop 73, 104
Shackleton, Ernest 148
Shadbolt, Maurice 14
Sharp, William 171
Shaw, Peter 13
Sheraton Auckland 69
Shigeru Ban 13
Shortland Street, Auckland 54, 76
Siberia, Manawatu 15
Skinner, William 77
Smith, Alfred 81–82
Solomon, King 90
Southland Hospital Board 181
Southland Museum and Art Gallery 181
Square House, New Plymouth 99
St Andrews Church, Auckland 82
St Barnabas Point 75
St Benedict's Church, Newton 12
St George's Hospital, Christchurch 139
St James Theatre, Auckland 14
St James', Brightlingsea 73
St John the Baptist Church, Christchurch 13
St John's Cathedral, Napier 8, 102–109
St Mary's Procathedral, Auckland 12
St Matthew's Church, Auckland 38
St Matthew's Church, Dunedin 77
St Paul's Church, Auckland 62, 70–77
St Paul's Church, Dunedin 160
St Paul's, Oamaru 77
St Paul's, Napier 109
Stacpoole, John 56, 64, 86, 128, 154
Stafford, Edward 128
State Opera House, Wellington 50
Steele, Louis J. 83–84
Stenberg, Ron 86
Strand Arcade, Queen Street 44
Strathmore Boarding House 101
Sunnyside Mental Hospital, Christchurch 157
Supreme Court, Auckland 82, 85

T
T.J. Edmonds Ltd 134–141
Tamati Waka Nene 21, 22
Tararua Range 114
Taylor, E. Mervyn 86
Te Arawa 115
Te Ati Awa 112
Te Aute College 109
Te Kawa a Maui 95
Te Kooti (Arikirangi Te Turuki) 90
Te Rauparaha 112
Thatcher, Frederick 181
The Gables, New Plymouth 181
The Pah, Hillsborough 56
The Rocks, Sydney 10
Theatre Royal, Christchurch 50
Theosophical Society Building, Christchurch 140–141
Thomson, John Turnbull 165
Threadneedle Street, London 80
Treaty House, Waitangi 8
trench warfare 21
Truby King Recreation Reserve 157
Tuhoe 90, 94
tukutuku 116
Turnbull, Thomas 122
Turner, Dennis K. 86
Twiss, Greer 86

U
universal penny postage 170
University of Auckland 57
Urewera 90
   Hiona 8, 88–95

V
Vaile, E. Earle 68
Victoria Arcade, Auckland 10, 44, 58, 62, 78–87
Victoria Flour Mills and Steam Biscuit Factory 64
Victoria, Queen 65, 117, 176
von Meyern, Ellen 44

W
Wachner Place, Invercargill 173
Wade, Norman and Henry 85
Waiapu 104–105, 108
Waihou (Thames) 112
Waikanae 112, 115
Waikare River 95
Waikawa River 114
Waimana 94
Waimate North 112
Wairau, Marlborough 115
Waitangi Treaty House 100
Waitara, Taranaki 115
Wakatipu, Lake 176
Ward, Joseph 170, 172, 179
Warren, Philip 50
Waterloo Quadrant, Auckland 33
Watkins, Kennett 83
Waverley Hotel, Auckland 12, 80
Weeks, John 86
Wellington
   Parliament Buildings 8, 92, 118–125
Wellington Observatory 147
Wellington Opera House 51
Wellington Provincial Government 120
Wellington Town Hall 125
White, Henry 62
White, Minnie 84
Wild, Adam 39
Wildman's bookshop, Auckland 84
Wilkinson, James 66
Williams Memorial Chapel 107
Williams, Bishop William 104
Williams, Captain 144
Williams, Henry 72
Williams, Reverend Samuel 112–113
Williams. H.B. 48
Williamson, James 56
Willow Field, New Plymouth 99
Wilson, Rear-Admiral 28
Windmill Road, Auckland 62
Winkelmann, Henry 85
Wiseman, William 28
Woolston, Christchurch 136
Worrall Jewellers Ltd., Auckland 86
Wright, Frank and Walter 83–84

PICTURE CREDITS

Alexander Turnbull Library, Wellington New Zealand:
p.16–17 [A-079-007]; p.20 [E-320-f-003]; p.23 [A-079-030]; p.102–103 [1/1-002978-G]; p.105 [1/2-001384-G]; p.110–111 [1/2-046157-G]; p.113 [D-010-002]; p.118–119 [1/2-106917-F]; p.121 [G-520]; p.124 [1/2-019528-F]; p.126–127 [1/1-011257-G]; p.155, front cover (top) [PAColl-8769-07]; p.166–167 [1/2-022524-F]

Archives New Zealand:
p.153 [DAHI/20271/D266/520d]

Auckland Art Gallery Toi o Tamaki:
p.73 [1937/15/25, gift of Harry Kinder, 1937]; p.78–79 [1986/5, gift of the P A Edmiston Trust, 1986]; p.85 [1983/11/26, gift of John Stacpoole and John fields, 1983]

Auckland War Memorial Museum:
p.4 [RMS-NZA9. Morrison, Robin (1991)]; p.34–35 [RMS-NZA88-d. Morrison, Robin (1991)]; p.40–41 [C6005]; p.43 [RMS-ACS44. Morrison, Robin (1982)]; p.57 [A563]; p.81 [B9915. Lediard, Reginald Silvester]; p.91 [C5884. Bourne, George (1908)]; p.93 [C5880. Bourne, George (1908)]; p.141, back cover (left) [RMS/FTR64/92. Morrison, Robin (1979)];

Barfoot & Thompson Ltd:
p.37

Christchurch City Libraries:
p.179 [CCL-PhotoCD04-IMG0048]

Fairfax Media/The Press:
p.149

Goodman Fielder New Zealand Limited:
p.137

Hanly, Gil:
p.49

Hocken Collections, Uare Taoka o Hakena, University of Otago:
p.158–159 [S11-154b]; p.161 [S11-154c]; p.174–175 [S13-143a]

Kete Christchurch:
p.138

Lincoln, Mark:
p.142–143; p.145, back cover (middle)

Macmillan Brown Library, University of Canterbury:
p.134–135 [IPD# 25519]

Museum of New Zealand Te Papa Tongarewa:
p.177 [C.015131]

Nature's Pic Images, Rob Suisted:
p.19 [15225TN20]

Puke Ariki, New Plymouth:
p.96–97 [A66.060]; p.99 [A65.921]

Sir George Grey Special Collections, Auckland Libraries:
p.6 [7-A4496]; p.11 [1-W627]; p.26–27 [1-W1409]; p.29 [4-2482]; p.32 [1-W4]; p.45 [7-A11765]; p.46 [7-A11766]; p.52–53 [4-680]; p.55 [1-W1236]; p.60–61 [1-W490]; p.63 [7-A5028]; p.66 [4-8560]; p.70–71 [1-W1692]; p.75 [4-1]; p.87 [4-259]; p.88–89, front cover (bottom) [7-A3323]; p.108 [AWNS-19310225-44-2]; p.169 [AWNS-19371208-45-2]; p.172 [35-R621]; back cover (right) [7-C81]

State Library of Victoria:
p.150–151 [IAN26/11/84/196]

The Nelson Provincial Museum:
p.129 [C941]; p.132 [C15161]